I0052789

Manage Projects with Purpose

And the Goals Will Take Care of Themselves

James J. Allen, Jr.

Response Resource Publication

Manage Projects with Purpose

And the Goals Will Take Care of Themselves

Response Resource Publication

First Edition April 2024

ISBN:

979 8 9905191 0 7 Paperback
979 8 9905191 1 4 Digital Online
979 8 9905191 2 1 Hardcover

Published in the United States by

Response Resource Publication

Herndon, Virginia 20171

ResponseResource.com

Table of Contents

INTRODUCTION 1

CHAPTER 1 - CHOOSING AN APPROACH TO PROJECT MANAGEMENT............. 7
GOAL ORIENTED PROJECT MANAGEMENT 11
PURPOSE-DRIVEN PROJECT MANAGEMENT 15
The Purpose of a Purpose-Driven approach............................. 25
Purpose is More Than Meeting Requirements........................... 30

CHAPTER 2 - MOTIVATION 35
SELF-ESTEEM 36
SELF-ACTUALIZATION 37
WHAT MOTIVATES CLIENTS? 37
LINKEDIN COMMUNITY VIEWS... 42

CHAPTER 3 - LEADERSHIP AND ETHICS IN PROJECT MANAGEMENT 43
LEFT BRAIN VS. RIGHT BRAIN: FACT OR FICTION...................... 46
GOAL ORIENTED THINKING... 49
PURPOSE-DRIVEN THINKING 50

CHAPTER 4 – BALANCING PURPOSE AND GOALS 53
WHAT ARE THE CHARACTERISTICS OF PURPOSE? 53
WHAT ARE THE CHARACTERISTICS OF GOAL DRIVEN? 54

CHAPTER 5 - TIME FOR A RE-COMPETE 57
VULNERABILITY... 57
Performance.. 58
People and Staffing 59
Pricing 61
Arrogance.. 63
Capture and Proposal Process....................................... 64
REPEL THE ATTACKERS 67
Performance.. 68
People and Staffing 70
Pricing 71
Arrogance.. 73
Capture and Proposal Process....................................... 74
DON'T KNOW TOO MUCH... 75

CHAPTER 6 – CLIENT ENGAGEMENT PROCESS................................79
SHAPING AND TRACKING RELATIONSHIPS................................. 84

Do your homework! *84*
Prepare for the engagement. *85*
Engage *85*
Document the engagement. *89*
EVALUATE RESULTS 89

CHAPTER 7 - CAPTURE MANAGERS **91**
HARDEST JOB 91
RESPONSIBILITIES 96

CHAPTER 8 - WHAT IS VALUE? **105**
CREATING VALUE PROPOSITIONS 106
MULTI-ATTRIBUTE COMPETITIVE ANALYSIS 115
Are you competitive? *115*
Remember Teaming Can Build Strength *118*

CHAPTER 9 - CONGRATULATIONS, YOU WON! **121**
UNDERSTAND THE PROJECT VISION! 124
VALIDATE YOUR BASIS OF ESTIMATE (AND REBALANCE WHERE POSSIBLE) 133

CHAPTER 10 - IMPACT OF SELECTING PURPOSE-DRIVEN APPROACH **149**
NEEDS ASSESSMENT 152
CHOOSE A MANAGEMENT STYLE 155

CHAPTER 11 - SCOPE WILL CREEP **159**
KEEP THE CONTRACTS' SHOP INFORMED 160
BENEFITS ARE EVOLUTIONARY 162
EMBRACE FLEXIBILITY 164
ITERATIVE DEVELOPMENT APPROACH 166
REQUIREMENTS AND BENEFITS CONTINUE TO MORPH. 172

CHAPTER 12 - FIVE TYPES OF REQUIREMENTS **177**

CHAPTER 13 - CONCEPT VISUALIZATION **187**

CHAPTER 14 - THE ULTIMATE STRUGGLE **191**

CHAPTER 15 - CONCLUSION **197**

REFERENCES **201**
BOOKS: 201
ARTICLES: 204

GLOSSARY OF TERMS **207**

APPENDIX A – SAMPLE EVALUATION FACTORS..**209**

APPENDIX B – CEP COURSE SUMMARY...**221**

APPENDIX C – VALUE PROPOSITION **225**

APPENDIX D – SELECTED DODAF VIEWS...**227**

APPENDIX E – BUILDERS – AKA PROJECT MANAGERS **233**

About the author:

Jim Allen is a skilled Program Executive with extensive experience working with the Intelligence Community, the Department of Defense, Civilian Agencies and Commercial accounts. He applies a disciplined engineering approach to all new and expanded opportunities.

He provides significant expertise in Project Management, P&L Leadership, Opportunity Shaping, and Solution Architecture. He has applied this expertise to the design and implementation of cost-effective solutions for complex problems addressing customer issues for both Government and Commercial accounts. Over the past five decades, he has learned that you have two choices when pursuing and executing projects, either you focus on

The Client Wining or You Wining.

If you plan on making project management a long-term career, it is best to ensure that the "Client Wins". This is true whether the client is internal or external to your organization. If you let client success be your guiding principle, the goals will take care of themselves. In earlier phases of his career, he wanted to manage projects correctly and by the book. He wanted to do the job the right way, on time and on budget. He later learned an invaluable lesson,

Do Right by the Client
And the Goals Will Take Care of Themselves.

Introduction

Over the past five decades, I have learned that you have two choices when pursuing and executing projects

Client Wins or You Win.

If you plan on making project management a long-term career, it is best to ensure that the "Client Wins". This is true whether the client is internal or external to your organization. If you let client success be your guiding principle, the goals will take care of themselves.

In a recent issue in the Washington Post (March 2024) there was an article about a purpose-driven medical provider being acquired by a larger corporate entity.

The medical provider had a purpose-driven approach. *"If you spend more on a patient upfront, you will produce savings in the long run."* She went so far as to pay for a patient's cab ride to the office when the patient needed to come in. A cab ride might cost the provider $10, but an ambulance ride following a 911 call will cost the patient a lot more.

Her philosophy was that even though transportation was expensive, having patients in the hospital was even more expensive. The provider focused on ***doing right by the patient.*** She focused on ensuring that her ***Client Wins.***

This principle created an admirable level of trust and loyalty. Things changed after the practice was acquired by the larger organization. The ***"doing right by the patient"*** philosophy eroded.

The acquiring company began to institute changes that they felt positioned the provider in a more sustainable financial posture. The changes were more focused on profitability in the short term. This has become more of a ***Company Wins*** philosophy.

One Medical doctor, who spoke on the condition of anonymity to protect his job, said he was concerned that a new operations role is "an easy answer to any profitability question [that] can pretty rapidly turn a well-paced and humane job to a factory-style rat race."

Patients began to express concern over the redirection and over the loss of the patient-centric philosophy. Leadership often pushes project managers into a ***factory-style rat race***, forgetting to realize the underlying needs of the client.

In earlier phases of my career, I focused on the bottom line, not client needs. I entered the rat race. I wanted to manage my projects correctly and by the book. I wanted to do the job the right way, on time and on budget. I later learned an invaluable lesson:

Do Right by the Client

I have seen a major trend in the performance of program and project managers. I believe that the number one change that we needed to make was to spend more time being purpose-driven, more time focusing on client needs.

We need to identify and talk with stakeholders to ensure that whatever solution that we are developing meets the client's needs and provides value for them.

I also learned that needs evolve over the life of a project. There will be an initial set at inception. Think of that phase as the pursuit phase of a project. As the project matures there will be a better understanding of client needs. As you build the business case and ask the "why?" of the project, more needs will surface. Even after the initial build, there will be more.

> *"It's unbelievable how much you don't know about the game you've been playing all your life."*
>
> Mickey Mantle

Tom Roden and Ben Williams in *"Fifty Quick Ideas"* discuss that you are likely to find more needs as you perform a retrospective on the project. The perspective of Purpose-Driven needs assessment is never ending.

For the past few years, I have been developing and conducting purpose-driven training sessions. In those courses, I have attempted to emphasize the need for collaborating with stakeholders. If you are to be purpose-driven, you must collaborate and adapt. If you do that, *the goals will take care of themselves.*

The purpose of this book is as a reference text that provides readers with a quick reference on the topic.

I share some of the observations and lessons that I have learned while both managing projects and working with hundreds of students. Over the past five decades, I have experienced both the right and wrong ways to manage projects.

"I have not failed; I've just found 10,000 ways that won't work."

Thomas Edison

I share some stories as well as these lessons learned. I will touch on projects in three states: pursuit, execution, and retrospective. Hopefully this will motivate you to be a better purpose-driven project manager.

Some stories relate to pursuing a contract. Pursuit of an opportunity is a project and should be handled like all other projects. The difference is that the end state is a win or a loss, not a delivered solution.

Hopefully pursuits will become wins that result in execution. Some stories relate to ongoing projects. Most notably, projects have the tendency to exhibit scope creep. Almost any large project will experience scope creep that requires that we take a step back in the business process loop, gathering more needs and more requirements before proceeding.

Other ongoing activities also need to be revisited.

"Remember, the reasons for rejecting something based on previous experience may no longer be valid."

Tom Roden

Most projects benefit from a healthy retrospective to validate the continuation of a need or to identify new needs.

Clients Never Run Out of Needs.

*The wise man does not lay up
treasure. The more he gives to
others, the more he has for his own.*

Lao-Tse

Chapter 1 - Choosing an Approach to Project Management

As you form your program management office, PMO, you should settle on selecting an approach to manage. Are you going to engage with the client or simply report on progress. The latter is much easier, and most PMs use that approach.

If we are to satisfy client needs, we must understand the purpose of the project, not just the requirements.

In 2019, Lisa Perrine, Ed.D., authored an article, "Purpose-Driven Projects: Start with the Why".

The concept of "purpose" is the real reason behind an event, activation, or technology purchase. But for too long, technology professionals have been hesitant to ask customers, "what's your project's purpose?" We may ask how they're going to use a room, a piece of equipment, or a user interface. But how often do we truly understand the business rationale for our clients' technology investments? The "why?"

Let's start by looking at how PMI defines a project.

"A project is a temporary endeavor undertaken to create a unique product, service, or result."

Projects are undertaken to fulfill objectives by producing deliverables, a purpose to be achieved, a result to be obtained, a product to be produced, or a service to be performed.

Achieving purpose is the key, usually overlooked component. Where does purpose fit in a stagnant set of goals/requirements. A good PM searches for purpose by collaborating with stakeholders.

Dr. Perrine goes on to explain why we have hurdles that we must clear up when working with stakeholders.

Hurdle One – "Because I Said So"

Hurdle Two – "It's Too Late"

Hurdle Three – "Not in My Contract"

As we work with our clients, we must understand the difference between "why" (the real purpose) and goals and objectives.

Mathew Emmanuel Pineda published an interesting on-line article on the Profulus.com website in October 2020. He provided a simple definition of the difference between purpose and goals.

A statement of purpose and a description of goals and objectives are important elements that are usually found in specific types of result-oriented documentations and manuscripts such as strategic plans and scholarly publications. Thus, these three concepts are essential in undertakings that involve planning or research and inquiry.

Knowing the difference between purpose, goals, and objectives is essential not only in defining the scope and limits of a particular undertaking but also in appreciating and achieving desired results.

He provided the following simple example for personal development.

Purpose - *Create a life that is driven by the need to regularly seek a sense of fulfillment and accomplishment.*

Goal – *Establish a profitable business venture.*

Subordinate Objectives -

Save money to raise capital.

Take short courses in entrepreneurship and business management.

Build a network with potential partners and clients.

We must do a similar decomposition when we work to manage a project. But first we must understand the purpose - the value that the client expects to receive.

PMI has provided a vast body of knowledge that helps us navigate goals and objectives. The tools that we are provided however, fall short in answering the purpose question:

When you undertake a project, we must keep in mind that we must create value. In Mathew's personal development project example, it was straightforward. "Seek a sense of fulfillment and accomplishment."

Joey Reiman in his book "The Story of Purpose" makes an important observation:

He begins by mentioning that when it's all about the profits, you're going to fall into trouble. He talks about how people who focus on the bottom line are in a race to the bottom. The mission to get rich is not sustainable without a guiding sense of purpose, which provides energy, partners, and the team of people who allow you to be successful.

Our purpose as project managers should be to *bring value to the client, to understand and fulfill his/her needs.*

You also need to understand the approach that you wish to follow to create that value. We need to understand "why?".

Remember to ask these questions:
- Who is the client?
- Does the client know us?
- What does the client state regarding his wants?
- What does the client infer he wants?
- What does the client value?
- How are we going to give the client what he wants?

- How can we prove that we can deliver?
- How is the client going to benefit. (WIIFM?)

WIIFM – What is in it for me (the client)?

Most of us have been trained to treat project management as a goal-oriented process. We need to set goals for time and budget. We need to deliver the product and never ask "why?".

Steven Krupp and Paul Schoemaker in their book, "Winning the Long Game" discuss this issue.

"Division leaders tend to fucus on short-term unit goals, not long-term threats or opportunities".

"The key is to open your eyes to where your customers are heading, ideally before they fully realize it themselves. As Henry Ford said, 'If I had asked people what they wanted, they would have said faster horses."

A goal-oriented process will get the job done, but is it sufficient to meet the customer's needs?

Goal Oriented Project Management

"The purpose of project management is to foresee or predict as many dangers and problems as possible; and to plan, organize and control activities so that the project is completed as successfully as possible in spite of all the risks."

Burek, Paul (2008). Creating clear project requirements: differentiating "what" from "how" Paper presented at PMI® Global Congress 2008

How many of you believe that this is a good purpose statement for Program and Project managers? If you follow this principle, you are likely to win in the eyes of your management. What about the client? Are we addressing his or her evolving needs?

Does completing successfully include delivering meaningful value or are we just meeting the goals set in the project charter?

Most program managers are taught to be goal oriented. They need to hit a budget, meet the schedule, and stay within the scope provided by the client.

If you look at earlier versions of the PMBOK, the concept of pre-project work was to produce a business case. The business case was then turned into a charter and the charter into a program management plan (PMP.)

What is missing is the concentration on purpose throughout the project lifecycle.

We assumed that we had a solid representation of the stakeholders' needs and the benefits that they expected to receive. We could then put our heads down and go to work.

Current Project Life Cycle Remains Somewhat Linear

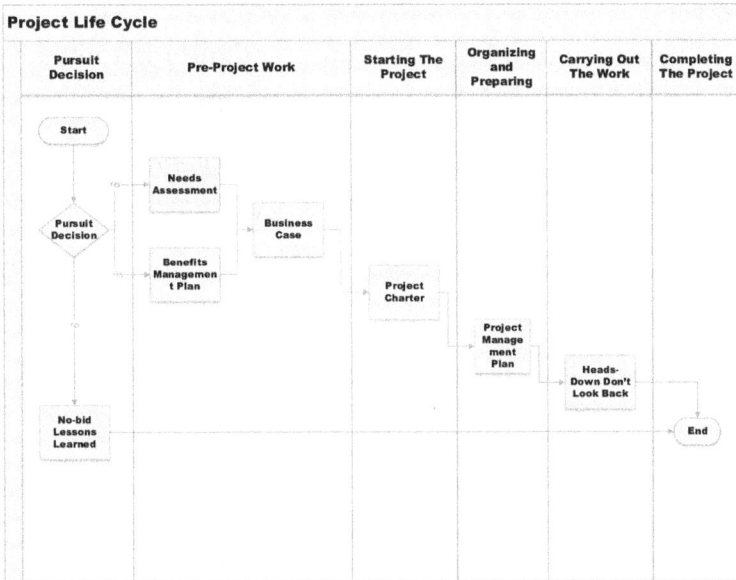

Project Life Cycle

Pursuit Decision	Pre-Project Work	Starting The Project	Organizing and Preparing	Carrying Out The Work	Completing The Project
Start					
	Needs Assessment				
Pursuit Decision	Business Case				
	Benefits Management Plan	Project Charter			
			Project Management Plan		
				Heads-Down Don't Look Back	
No-bid Lessons Learned					End

Works Well for Predictive Workflow

Did the client take all the steps in the workflow? Did the client fully understand the "why" for doing the project? Has a complete needs assessment or benefit plan been completed before you start work? Should you have this information if you are going to deliver meaningful solutions?

Does the Sponsor Always Communicate the Purpose?

"The WHY of the Project"

The project sponsor is generally accountable for the development and maintenance of the project business case document. A needs assessment and benefits plan should precede the business case.

As we often find, development and maintenance of the project's benefits plan become an iterative activity.

"Regardless of the business analysis approach used, elicitation and analysis are usually iterative."

PMI Business Analysis for Practitioners

Determining the why of the project is an iterative process. PMs must learn to illicit the whys from the stakeholders.

"On large programs, there are many stakeholders. In addition, in considerable challenge of maintaining alignment and expectations that will inevitably diverge (and may not have been in perfect agreement to begin with), you also need to deal with changes as the roles of specific individuals shift, the focus of your program evolves, and the people involved with your program come and go."

Tom Kendrick, PMP

As we have learned in the Agile process, each time we build another build we learn more about the problem that customers are trying to face. How do we handle it?

Success is defined by satisfying the "why" behind the requirements.

Purpose-Driven Project Management

If you are purpose-driven, you are putting the emphasis on the client and trying to make sure you understand the root causes of the pain that the client is facing.

What do you think about this definition of purpose-driven project management? With this definition, you address the underlying client needs. You address the "Why?" of the project.

The purpose of project management is to make the customer successful by identifying the underlying needs of the client and satisfying those root causes. (Success is not about just meeting the requirements, it's about discovering and fulfilling the "why" behind the requirement.)

David Allen in "Getting Things Done" discusses the value of thinking "WHY".

The Value of Thinking About Why

Some of the benefits of asking why:

- *It defines success.*
- *It creates decision-making criteria.*
- *It aligns resources.*

- *It motivates.*
- *It clarifies focus.*
- *It expands options.*

People love to win. If you're not totally clear about the purpose of what you are doing, you have no game to win.

Success is better defined as resolving the pains experienced by the client, even if they were not able to communicate properly. It is necessary that you understand the why (Purpose) of the project if you are going to build a solution that is valued by the client.

Over the past decades, the Program Management Institute has shifted from a goal-oriented point of view to one of stakeholder focused (Purpose-driven) program management.

Version 2 of the PMBOK (the version I originally studied) really did not address stakeholders very much. Some material was added in version 3 that began to address stakeholders. It was oriented toward reporting information to stakeholders, not gathering information. Managing stakeholder expectations were added in version 4. Now in version 5 and 6 of the PMBOK, a whole new chapter dealing with stakeholder engagement was added.

The Why (Purpose) of a Project Has Grown in Importance

- *Determine information and communication needs of stakeholders (PMBOK version 2.0)*
- *Manage stakeholders (PMBOK version 3.0)*
- *Project Stakeholder Process - split from overall project communication – Effectively engage stakeholders throughout the project life cycle, based on the analysis of their needs, interests, and potential impact on progress success. (PMBOK version 5.0)*
- *Benefit Management – A project benefit is defined as an outcome of actions, behaviors, products, services, or results that provide value to the sponsoring organization as well as the project's identified beneficiaries. The development and maintenance of the project's benefits plan is an iterative activity. (PMBOK 6.0)*

Program Management Body of Knowledge

The problem that we face is that we now know that we should be engaging with stakeholders but there is extraordinarily little material telling us how we should engage. We are process oriented and we need a process to follow to achieve the desired results.

The Agile Manifesto provides further emphasis. The Manifesto now requires that our highest priority is to satisfy the customer through early continuous delivery of valuable software while welcoming change requirements.

Current Best Practices Require Understanding of "Why?"	
Four Values of the Agile Manifesto	**Principles That Drive Purpose-Driven Projects**
• Individuals and interactions	• Our highest priority is to satisfy the customer through early and continuous delivery of valuable software.
• Working software	• Welcome changing requirements, even late in development. Agile processes harness change for the customer's competitive advantage.
• Customer collaboration	• Businesspeople and developers must work together daily throughout the project.
• Responding to change	• Embrace change

Change, scope creep and dissatisfaction are ways that stakeholders try to communicate their pain. We must pursue the pain if we are to deliver solutions that truly resolve customer pain and deliver value.

- We must not avoid change requests.
- We should welcome scope creep.
- We should examine stakeholder dissatisfaction.

A purpose-driven value management view of program/project management requires a closer relationship with stakeholders than the ones we used to have in the triple constraint world of schedule, scope, and budget (C3).

New Role of the New Project Managers		
	C3* Paradigm	PVM Paradigm
Paradigm Focus	Satisfy Predefined Objectives	Maximize Project Value
Paradigm Perspective	Operational	Strategic
Project Manager's Responsibility	Satisfy Requirements	Create Value
Project Definition	Define Requirements	Identify Customer Needs
Project Planning	Baseline	Centerline
Project Execution	Minimize Variation	Identify Opportunities
Project	Project	Project

Manager's Role	Administrator	Leader
*C3 – (time and cost) Paradigm mainly focuses the management of projects on planning and conformance with schedule, budget, and scope during the implementation of projects. PVM focuses on creating value.		

What do we mean by Baseline vs Centerline? Baselines imply rigidity while centerlines infer guidance as we move down the road toward a solution.

How do you feel about Minimize Variation vs Identify Opportunities?

Haven't we been taught constraint? Build the plan, stick to the plan, deliver. Is it strange to look for additional opportunities to help customers resolve their pain?

Oh yes, Satisfy Requirements vs Create Value? What does that mean and how do you do it?

What we needed to do was to change this process to some semi waterfall where we get all the facts up front and put our heads down and carry out the work.

We needed to change to one that is more iterative doing requirements analysis design iteratively each time until we finally delivered the final project. We need to constantly look for solutions that resolve stakeholder pain.

We need to move away from the traditional waterfall...

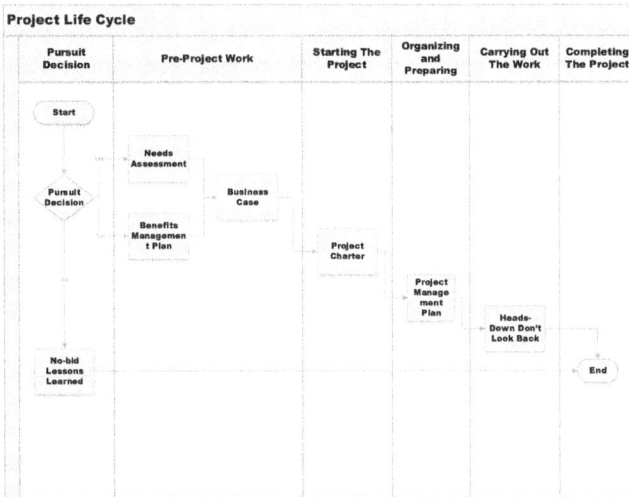

Project Life Cycle

Pursuit Decision	Pre-Project Work	Starting The Project	Organizing and Preparing	Carrying Out The Work	Completing The Project

Start

Needs Assessment

Pursuit Decision

Business Case

Benefits Management Plan

Project Charter

Project Management Plan

Heads-Down Don't Look Back

No-bid Lessons Learned

End

...to an iterative agile based approach.

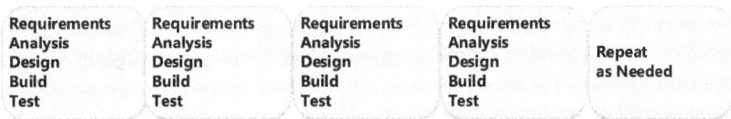

Requirements Analysis Design Build Test	Requirements Analysis Design Build Test	Requirements Analysis Design Build Test	Requirements Analysis Design Build Test	Repeat as Needed

Iteration-Based Agile
"Agile Practice Guide"

If a project is very predictive and you do not have much anticipated change in a project, you can build using a waterfall or one or two release approach.

The requirements definition process begins with the elicitation of stakeholder requirements, the first step of which is to identify the stakeholders from whom those requirements are to be gathered.

Michel J. Ryan (INCOSE)

We start by decomposing the primary requirements into sub and sub-sub requirements until we have sufficient granularity necessary for the build.

Decomposition of the initial requirements is just one of the steps.

If a project is very predictive, you do not have much change in a project you can build using a waterfall or one or two release approach.

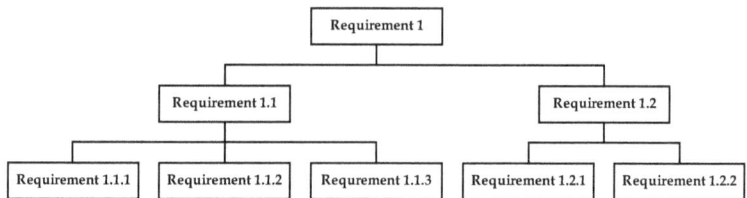

Traditional Hierarchical Decomposition of Requirements

You just need to address each of the sub-requirements and make sure that you build tests to ensure that they have each been addressed.

When the frequency of delivery goes up or the degree of change becomes greater, you must move to a fully agile process where the customer is involved on a day-to-day basis to tell you what you're doing right and what you're doing wrong.

Most projects are hybrids. Some predictive components such as foundational infrastructure are best implemented incrementally. While more user centric services are best implemented iteratively.

When the frequency of delivery goes up or the degree of change becomes greater, you must move to a fully agile process where the customer is involved on a day-to-day basis to tell you what you're doing right and what you're doing wrong.

The Continuum of Life Cycles

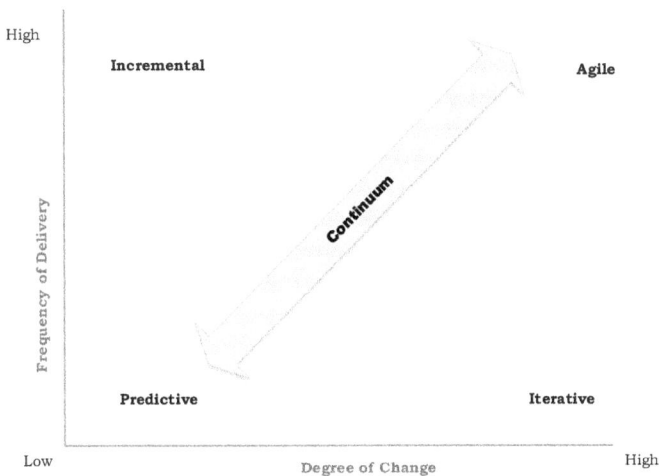

Dr. Ryan makes an interesting observation. "In requirements engineering, a stakeholder is commonly defined as someone who has a stake in the project — that is, someone who is affected by the system in some way or can affect the system in some way."

Their input to the system lifecycle is crucial— careful selection of appropriate stakeholders is therefore fundamental to the success of the project.

We need to use care in selecting and authorizing stakeholders. Dr. Ryan proposes an alternative definition.

> *A stakeholder is someone who has a right to influence the system.*

This concept becomes more evident when we discuss the need to keep contracts involved when validating requirements. The stated requirement may be real, but if the stakeholder is not authorized to establish a requirement, we might not be able to fulfill his request.

Understanding the Need to Know "Why" is Not New.

> *"He who has a "why" to live for can bear almost any how."*
> Friedrich Nietzsche, *"Twilight of the idols"* 1888

> *"Until thought is linked with purpose there is no intelligent accomplishment."*
> James Allen, *"As a Man Thinketh"* circa 1903

> *"He [who] knows the "why" for his existence... will be able to bear almost any 'how'."*
> Viktor Frankl, *"Man's Search for Meaning"* 1959

"Until we understand the 'why' of a requirement, we do not understand the requirement."

Jim Allen, circa 2018

The Purpose of a Purpose-Driven approach

The effect of a purpose-driven culture was quantified in a Korn Ferry study entitled "*People on a Mission*." The study reported that 90% of people who worked in a purpose-driven organization said they felt engaged in their work. Further results showed that

"Companies with the proper focus posted compounded annual growth rates of 9.85% compared to a 2.4% for the whole S&P 500 Consumer Sector."

Elaine Dinos, principal of Korn Ferry's Global Consumer Market practice sums up the findings of the study in these words:

"When an organization has a clear purpose, it unleashes the power and drive of the entire workforce, harnessing and focusing that combined effort in one aligned direction."

Understanding the Concept of Why

If we are to be truly purpose-driven, we must understand the "why" of the project. We must keep our clients' objectives and outcomes in the forefront.

James Kruger, Executive Vice President of Capgemini Insurance puts it this way in an article that he published on LinkedIn on September 13, 2023:

We must be client focused. We must understand the "why" from the client's perspective. Understanding the concept of Why is not new. Once we understand the "why" it becomes much easier to understand the what and the how. It boils down to a simple concept, until we understand the why of a requirement, we really do not understand the requirement.

The "why" is the purpose for building the solution. We usually start a project with a given set of requirements and then simply decompose and build. A better approach may be ensuring that we understand the purpose, the "why". We should start with the Project Purpose Statement.

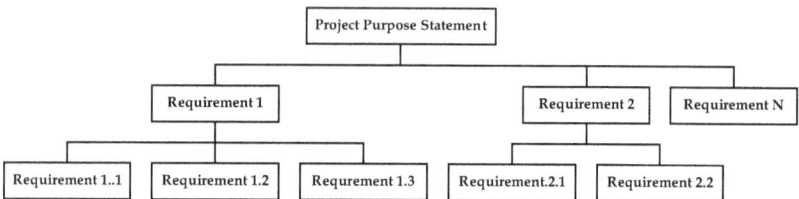

Decompose Starting with the Purpose Statement

Remember that the client sees things from a different perspective than the project manager.

The LinkedIn AI generate blog on this subject asks these wise questions:

- *Why are you doing this project?*
- *What value will it deliver to your organization or customers?*
- *Who are the key stakeholders and beneficiaries of the project?*

The project vision and purpose should be aligned with the answers to those questions.

Joe Pusz discussed these concepts in an article that he published in ProjectManagement.com. Project managers want to share how they are performing on their projects.

We want to show the client

How We are Doing.

The more detail the better.

We want to share how well we are managing the budget. We want to share our percentage of completion for each task. As project managers, we want to communicate and share our progress reports.

Does this satisfy the client? Client satisfaction is not a question that enters their minds, but should it?

The client wants to know if we are satisfying his purpose. It is great that you are following schedules and budgets but is it helpful to my bottom line?

Is the client realizing success?

National League

EAST	W	L	PCT	GB	HOME	ROAD	RS	RA	DIFF
*-Philadelphia	102	60	.630	-	52-29	50-31	713	529	+184
Atlanta	89	73	.549	13	47-34	42-39	641	605	+36
Washington	80	81	.497	21.5	44-36	36-45	624	643	-19
NY Mets	77	85	.475	25	34-47	43-38	718	742	-24
Florida	72	90	.444	30	31-47	41-43	625	702	-77
CENTRAL	W	L	PCT	GB	HOME	ROAD	RS	RA	DIFF
x-Milwaukee	96	66	.593	-	57-24	39-42	721	638	+83
y-St. Louis	90	72	.555	6	45-36	45-36	762	692	+70
Cincinnati	79	83	.488	17	42-39	37-44	735	720	+15
Pittsburgh	72	90	.444	24	36-45	36-45	610	712	-102
Chicago Cubs	71	91	.438	25	39-42	32-49	654	756	-102
Houston	56	106	.345	40	31-50	25-56	615	796	-181
WEST	W	L	PCT	GB	HOME	ROAD	RS	RA	DIFF
x-Arizona	94	68	.580	-	51-30	43-38	731	662	+69
San Francisco	86	76	.531	8	46-35	40-41	570	578	-8
LA Dodgers	82	79	.509	11.5	42-39	40-40	644	612	+32
Colorado	73	89	.451	21	38-43	35-46	735	774	-39
San Diego	71	91	.438	23	35-46	36-45	593	611	-18

Are We Achieving Purpose?

(What is our league standing?)

Eliciting purpose and understanding issues and challenges are a complex activities. The Business Analysis Playbook by PMI highlights several difficulties:

- *Conflicting viewpoints and needs among diverse types of users,*

- *Conflicting information and resulting requirements from different business units,*

- *Unstated or assumed information on the part of the stakeholders,*

- *Stakeholders who are resistant to change and may fail to cooperate and possibly sabotage the work,*

- *Inability to schedule time for interviewing or elicitation sessions because stakeholders cannot take time away from their work,*

- *Inability of stakeholders to express what they do or what they would like to do, and*

- *Inability of stakeholders to refrain from focusing on a solution.*

The elicitation process can be even more difficult.

- *Not able to gain access to stakeholders.*

- *Stakeholders do not know what they want.*

- *Stakeholders have difficulty defining requirements.*

- *Stakeholders do not provide sufficient detail.*

The problem that I am most often faced with is that the stakeholder comes to the elicitation process with a preconceived solution in his mind. The project manager should endeavour to redirect the discussion back to the "need", the purpose, not the perceived solution.

> *Purpose-driven project management* **is an approach that emphasizes aligning project goals and activities with a clear purpose or mission. Rather than merely focusing on completing tasks, purpose-driven project management considers the broader impact and significance of the project.**
>
> Microsoft Copilot AI

Purpose is More Than Meeting Requirements

In about 1984, Professor Kano produced a concept on how clients perceive our efforts.

He said that if you only do what the customer wants you to do, simply deliver the stated requirements as they were written in your obligation and are written in your project charter, they'll consider you to be a vanilla developer, a vanilla project manager.

You are doing what they are paying you to do, nothing more and nothing less.

Basic - Simply stated, these are the
requirements that the customers expect and
are taken for granted. When done well,
customers are just neutral but when done
poorly, customers are very dissatisfied.

Kano, 1984

Kano originally called these "Must-Be's" because they were the requirements that must be included and are the price of entry into the market.

You are delivering these BASIC requirements following a very goal-oriented approach to project management. Don't ask questions. Don't collaborate. Put your head down and build.

However, if you are to generate excitement, you must find problems the client did not even know they had and propose solutions for those problems. By doing so, you create value.

This added value adds excitement. It is realized as a delighter by the client, resulting in a client that is much happier with you and much happier with the resulting solution.

Excitement - Simply stated, these are
the requirements that are unexpected and
pleasant surprises or delights. These are the
innovations you bring into your offering.
They delight the customer when there, but
do not cause any dissatisfaction when

missing because the customer never
expected them in the first place.

Kano, 1984

Kano originally called these "Attractive or Delighters" because that's exactly what they do. We will talk more about Professor Kano in a later chapter.

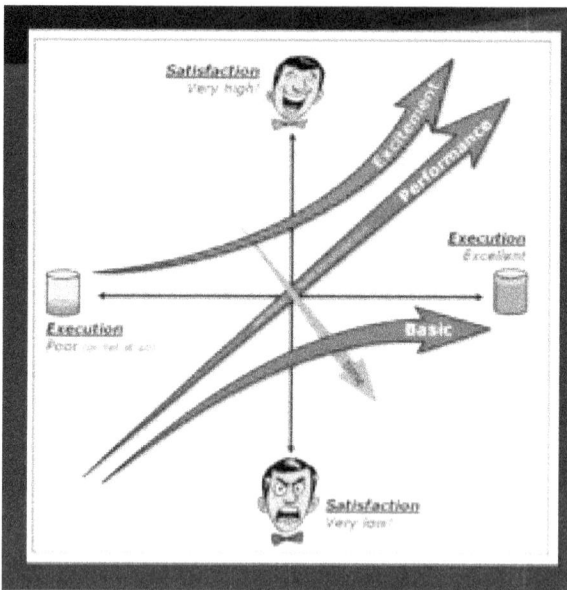

User Experience Magazine 1999

Remember, project managers are builders of solutions that create value, and meet the client's needs. "Appendix E – Builders" is a poem by Henry Wadsworth Longfellow that puts project management in the right perspective.

As a side benefit customers will come back to you and ask for more over the coming years.

You will also find this in "The Lean Product Playbook" by Dan Olsen. Dan goes on to say, "Discovery of Purpose Creates Satisfaction."

*Strive not to be a success, but rather
to be of value.*

Albert Einstein

Chapter 2 - Motivation

We are each motivated by our underlying needs. Maslow postulated the following: *"human needs are arranged in a hierarchy, with physiological (survival) needs at the bottom, and the more creative and intellectually oriented 'self-actualization' needs at the top."*

HIERARCHY OF NEEDS
BY ABRAHAM MASLOW

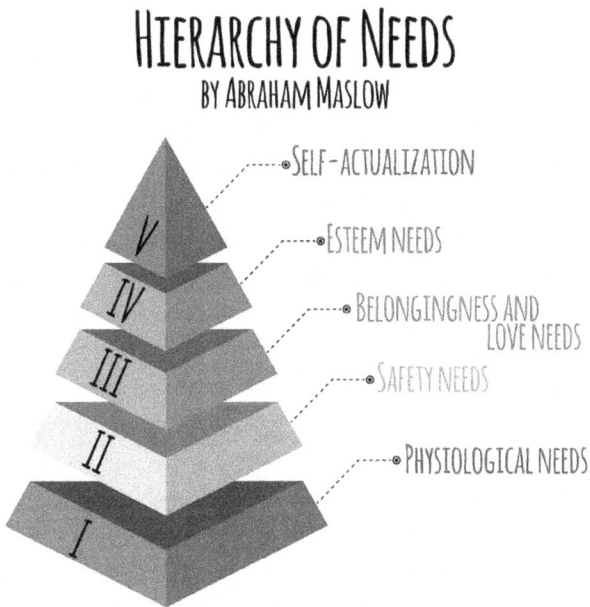

- SELF-ACTUALIZATION
- ESTEEM NEEDS
- BELONGINGNESS AND LOVE NEEDS
- SAFETY NEEDS
- PHYSIOLOGICAL NEEDS

Extracted for Maslow'S Hierarchy of Needs,by Saul Mcleod, Phd, Published January 24, 2024 in Simply Psychology

Let's assume that the bottom three needs are fundamental and probably not related to our project management job motivation.

However, Love and Belonging and Safety and Security may be client needs that need to be considered.

It is quite true that man lives by bread alone – when there is no bread. But what happens to man's desires when there is plenty of bread and when his belly is chronically filled?

Regarding the structure of his hierarchy, Maslow (1987) proposed that the order in hierarchy "is not nearly as rigid" as he implied in his earlier description. The bottom three often find their way into day-to-day work life, but let's concentrate on self-esteem and self-actualization. They are definite motivators in our project work life.

Let's look at the top two:

- How do they relate to project management?
- How do they relate to job satisfaction?
- How do they relate to client satisfaction?

Self-Esteem

We want respect from others. We want a sense of achievement in the work that we do. We want the self-confidence necessary to measure our self-esteem.

Esteem presents the typical human desire to be accepted and valued by others. People often engage in a profession or hobby to gain recognition. These activities give the person a sense of contribution or value.

Self-Actualization

Once we have self-esteem, we need to feel that we have a purpose. We need to understand that our inner potential is valued by others. We need to be creative. Don't forget, we want to be perceived as good people, doing the right thing.

We need to practice ways that motivate the stakeholders to uncover their underlying needs. We need to understand the "why".

What motivates clients?

Let's start by evaluating methods of persuading clients to work with us. Nick Kolenda in his book "Methods of Persuasion: How to use Psychology to Influence Human Behaviour", observes that

"First, offering any type of incentive will boost your persuasion, right? Wrong. Mounting research has disconfirmed the common dogma that all incentives lead to better performance. The main reason for that surprising discrepancy can be found in two types of motivation that result from different incentives: Intrinsic motivation – Motivation that emerges from a genuine personal desire (i.e., people perform a task because they find it interesting or enjoyable) Extrinsic motivation – Motivation that

emerges for external reasons (i.e., people perform a task to receive a corresponding reward)"

Nick postulates a seven-step process to effective persuasion:

1. *Mold Their Perception*
2. *Elicit Congruent Attitudes*
3. *Trigger Social Pressure*
4. *Habituate Your Message*
5. *Optimize Your Message*
6. *Drive Their Momentum*
7. *Sustain Their Compliance*

As you follow these steps, remember that incentives alone will not be sufficient to motivate.

- First, the client must be dissatisfied with their current state. Their status quo is just not working.
- Second, they need to be able to visualize the desired future state. That is part of our job as program and project managers. We must help them visualize the to be state where their dissatisfaction will no longer exist.
- Third, we must show them that we have the knowledge of how to get there.

With those three components, they will be properly motivated to work side by side with developers and side by side with the engineers to come up with a usable solution.

Motivation = (Dissatisfaction + Visualization + Knowledge)

Dissatisfaction (either internal or external)
Visualization of desired end or result
Knowledge of how to get there.

Client will only change (Change) if the pain that they are in, (PTI), is greater than the pain of making the change, (PC). Don't forget, Motivation must also be greater than the pain of making the change.

Client Will Only Change If (PTI>PC)

Motivation is also governed by purpose and value. Motivation can also be aspirational. **Price Waterhouse Cooper conducted a remarkably interesting study.** They found that 79% of business leaders believe that purpose is central to success but only 34% consider purpose in guiding their decision making. As PMs do we fall into that category, and do we share that characteristic? I hope not.

There's even further value in being a purpose in the workplace.

The employees are extremely happy when they can find meaning in day-to-day work. They find that when they solve problems for somebody else, they see meaning in their day-to-day work.

This is another interesting observation made by PwC.

Business leaders are more worried on the right side of the following chart. They are worried about reputation, distinction, and differentiation.

Value of Purpose in the Workplace

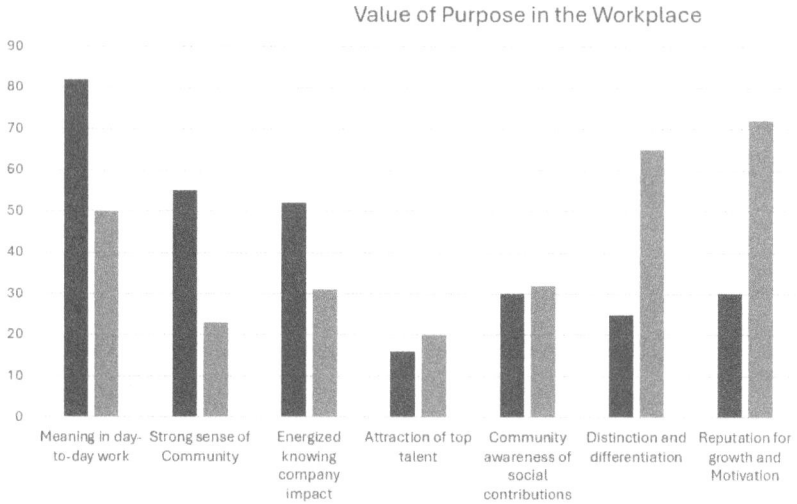

PWC's "Putting Purpose to Work Survey"

Employees, the people doing the work, are more concerned on the left side. They gain more satisfaction from hearing that the clients that they are working to help benefited from the work. This helps build self-esteem.

We must come up with a happy medium.

Everybody likes to say:

- This is why I've been doing this work.
- This is why I've put in those long hours.

- This is why I have been sweating and toiling.
- This is why I work to benefit others and gain their appreciation.

This is even more important than money. How do project and program managers live within this dichotomy. Why do project managers struggle with being purpose-driven project managers?

I believe fulfillment is a right and not a privilege. We are all entitled to wake up in the morning inspired to go to work, feel safe when we're there and return home fulfilled at the end of the day. Achieving that fulfillment starts with understanding exactly WHY we do what we do.

"Find Your Way", Simon Sinek

The main reason that we find ourselves in conflict is that project managers have been taught to be short-term, schedule focused. They tend to be process and goal oriented. They concentrate on doing things the right way. They generally don't ask "Why?"

"Employees most favor hearing about the impact of their company's products and services through client and customer stories, employee stories, and leadership messages."

PWC's "Putting Purpose to Work Survey"

LinkedIn Community Views

The LinkedIn community blog on program management asked the community about what it takes to stay motivated in program/project management. They have developed this first step:

1. Define your Purpose!

One of the first steps to maintain a positive attitude in program management is to define your purpose. Why are you doing what you are doing, and what are the goals and values that drive you? Having a clear sense of purpose can help you stay motivated, focused, and resilient in the face of challenges and setbacks. It can also help you communicate your vision and expectations to your team and stakeholders and align your actions with your desired outcomes.

The concept of WHY is a recurring theme.

Antonio Nieto from the Madrid Chapter of the Project Management Institute puts it this way:

"Our highest priority is to deliver projects better, to reduce the failure rate, to create more value for individuals and organizations, and to create more sustainable development in our economies and societies at large."

Chapter 3 - Leadership and Ethics in Project Management

When you undertake the role as a project manager, you must ask yourself - "what is my character?"; "what is my code of ethics?"; "why do I exist?" How do these factors affect my project management style?

> *"A PMO without Purpose manages Projects and Resources, but a Purpose-driven PMO Empowers."*

Joe Pusz
ProjectManagement.com

Are you interested in only the money that is going to come to you at the end of the week? Or are you driven to provide value to your client?

Ken Blanchard in "The Secret" presents a concept of self-evaluation that each project manager should ask themselves.

> *"Am I a serving leader or a self-serving leader?"*

The best PMs are serving leaders. Before you start a project, define WHY the PMO exists and how it will operate.

In a recent training session, the class determined that salespeople that were not strategic thinkers were referred to as that sleazy salesman. On the other hand, the modern salesperson or the modern program manager was more likely to reach back into his character, his foundation. and do the right thing for his clients. Can we say the same for project managers?

Should we do a self-assessment of our ethical and leadership characteristics? Does our client feel that we bring value to the relationship?

Leadership and ethics are about:

- Who you are as an individual.
- Your value to your client in the relationship
- Your character

Do we have a strong enough character to take on the role as a project manager?

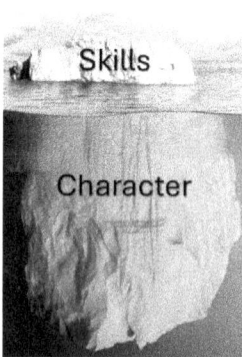

Character is built up by the principles and values that you learn from your family, your ethics that you've built around yourself in your company and your mission and purpose of what you're trying to do in life. It is your role and responsibility to your clients.

Many may say that the keys to project management are the skills that an individual exhibits. Look at the PMI exam. It is all about skills.

Blanchard makes another observation:

"As an organization, where possible we select people with both character and skills. But if we must choose between skills and character, we've made a fundamental decision on this issue:

We will select men and women of character and develop their skills."

Your principles, values and ethics anchor your day-to-day activities. If you follow them, you will be a much more effective project manager.

Project Managers Require a Strong Character

Your character is the sum of:

- Your principles and values
- Your ethics
- Your mission and purpose
- Your role and responsibility with clients

It seems that followers have a very clear picture of what they want and need from the most influential leaders in their lives: trust, compassion, stability, and hope.

"Strengths based Leadership" Tom Rath 2008

Your principles, values and ethics anchor your behavior in your Project Management role.

Left brain vs. right brain: Fact or Fiction

The left and right sides of the brain control different functions in the body. Some people believe that the left and right sides of the brain can determine personality and behavior. However, evidence to support this theory is limited.

In the 1960s, the neurobiologist Roger W. Sperry suggested that all people have one-half of their brain which is more dominant and determines their personality, thoughts, and behavior.

Due to the different functions of the two brain hemispheres, the idea that people can be left-brained and right-brained is tempting. Sharing the traits of both would be helpful in fulfilling the role as a project manager.

According to the dated theory, left-brained people are likely to be described using the following words.

- Analytical
- Logical
- Detail- and fact-oriented.
- Numerical
- Verbal

By contrast, the theory suggests that right-brained people are more creative. You might describe them as:

- Creative

- Free-thinking
- Able to see the big picture.
- Intuitive
- Likely to visualize more than think in words.

Right-brainers want to visualize the results before the solution is complete. They want to be able to say, "that is what I want when you finish the work." We will discuss this later when we address concept visualization.

Research suggests that the left brain vs. right brain theory is not correct. Having said that, we still tend to attribute our approach to life as either a left-brain or right-brain approach.

If we use the generally accepted notion that the left brain is focused on a short-term schedule focus, it would follow that left brain thinkers follow the rules and follow the process to build a solution following the correct approved methodology.

Left brainers would proceed in an orderly, logical, and analytical manner. Left brainers want to build the right way.

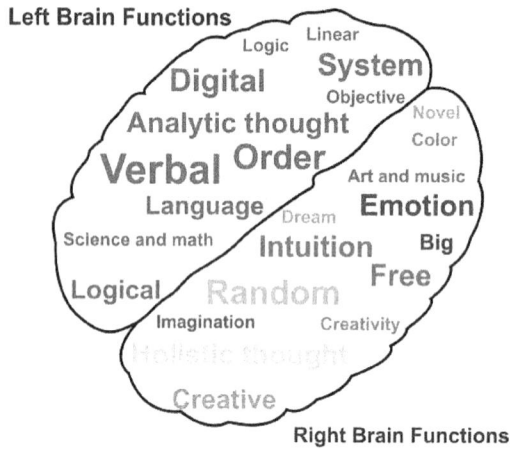

Left and Right Brain Functions

Left Brain Functions

Logic · Linear
Digital · System
Objective
Analytic thought · Novel · Color
Verbal · Order
Art and music
Language · Dream · Emotion
Science and math · Intuition · Big
Logical · Random · Free
Imagination · Creativity
Holistic thought
Creative

Right Brain Functions

Bottom line: We need to use both sides of our brain.

From the right brain perspective, we are more focused on building the underlined right solution.

Why Project Managers Struggle

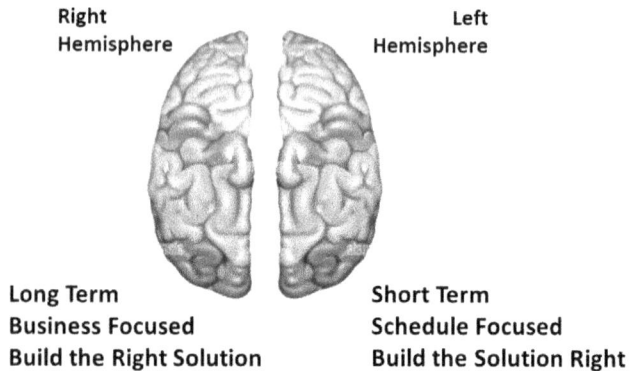

Right
Hemisphere

Left
Hemisphere

Long Term
Business Focused
Build the Right Solution

Short Term
Schedule Focused
Build the Solution Right

The brain is in conflict. The analytical, goal-oriented side wants to focus on technical metrics, schedule, budget, and measurable short-term aspects of the program. While the more creative, purpose-driven side wants to understand why we are doing what we are doing.

Goal Oriented Thinking

When you are goal oriented you tend to have a short-term point of view. You have a very finite set of options when you are goal oriented.

- The boundaries have been predefined for you. You can't color outside the lines.
- You can't exceed the budget.
- You can't go forward and ask for changes in scope to match what the client really needs.
- You can't take a little bit more time because the schedule is so tight and so focused.

Goal Oriented Projects Usually Depend on
Short-Term Tacitical Thinking.
Our Left Brain Focuses on Near Term.

- You are quantitative.
- You are internally focused.
- You operate from a position of scarcity.
- You are task oriented.

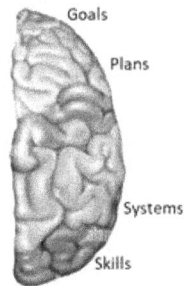

Goals
Plans
Systems
Skills

- You are short-term revenue focused.

What does it mean to be in a position of scarcity? Do we have a full understanding of the issues that we are addressing?

You have a very limited set of options when your goal oriented.

> *"Many projects have lengthy, technical, or deliverable-focused goals: a new software rollout, a new platform, an expansion program, a new set of company values, a reorganization, or a digital transformation project. Others use financial goals such as a 10% return on investment (ROI). Beware of overly complex goals and numerical targets such as these; they consistently fail to rally passionate commitment in a project."*
>
> **Harvard Business Review**
> **Antonio Nist-Rodriquez**
> **October 11, 2021**

The left brain focuses on these short-term objectives.

Purpose-Driven Thinking

Meanwhile, the right brain is worried about the principles of <u>doing the right thing</u> for people. It is about having the objective of being recognized as a contributor to the client's benefit.

Purpose-driven thinking is more qualitative in nature. It is more externally focused. You are not looking at yourself internally but looking at the effect that you have on other people.

You are less constrained. You have a position of abundance. Your boundaries are less constrained. You are open to change.

If your world is open to welcoming change, you feel free to look for the problems and look for the difficulties that clients are having.

Right Brain Tendancies

Principles
Missio

Purpos
Role

- Qualitative
- Externally Focused
- Position of Abundance
- Lead and Manage
- Creative
- Long Term Business Oriented

There is never a shortage of difficulties to identify. Yes, you must prioritize. You must find ways to fund. You must address these new problems. Just remember, you will never run out of problems to solve.

You want this customer to appreciate what you've done and you want to do more for them in the future.

Alleviating pain builds trust. You want to build trust. Trust is the secret of dealing with people. You also need to understand the leadership necessary to be effective. You need to understand yourself. You need to understand your true value to the client.

Purpose-Driven Projects Rely on Long Term Strategic Thinking

Our Right Brain Focuses on Strategic Concerns Building Trust

In a later chapter, we will dive a little deeper into the left/right brain concepts.

"Instead, weave purpose into the project's foundation. An effective purpose reflects the importance people attach to the project's work and reveals its fundamental reason for being. It awakens their intrinsic motivations and gets at the deeper WHY of the project beyond just making money."

Harvard Business Review
Antonio Nieto-Rodriques
October 11, 2021

Chapter 4 – Balancing Purpose and Goals

As you struggle between managing a project the right way and doing the right thing for the customer, you need to learn to balance the act.

Let's look at the characteristics of the approaches to each path.

What are the characteristics of Purpose?

- Long term external thinking.
- A selfless perspective.
- Qualitative and subjective.
- Service to others.
- The pride in our work every day.

Purpose is "The Why of Our Efforts."

Does that sound like the Pre version 5 of the PMBOK? I think not. Pre-Version-6 PMBOKs were focused on meeting management goals.

The right brain is focused on building the right solution. It has more of a long-term business perspective.

Remember the PWC study? People want to know that the work that they do is truly valued by the client.

It is not enough to deliver on time and on specs. Purpose-driven project managers want to deliver solutions that meet the needs of their clients.

Professor Kano went even further. He believed that it was our responsibility to help the client identify needs that they had not previously identified. Solving problems that they did not know were problems.

What are the Characteristics of Goal driven?

- They are specific measurable attainable realistic and timely.
- They are metric driven. What is my CPI?
- You ask yourself:
  - How am I doing against the schedule?
  - How can I know that I've done it on time and met the check points.
  - Have I delivered what I have been contracted to deliver?
  - Have I met the marks on my plan?
  - Have I delivered what I am supposed to deliver despite whether the client is achieving their desired results?

The Left Brain is Metric Driven

Managing a project from a goal-oriented perspective is a much simpler task. The left-brain is metric driven. It wants goals to hit and objectives that can be measured to achieve.

The goals should be **SMART**:

- **S**pecific
- **M**easurable
- **A**ttainable
- **R**ealistic
- **T**imely

Know your purpose, know your goals, Own both and communicate them with your stakeholders. This is a balancing act. Your stakeholders are a big part of the act.

One thing you should walk away with is - write this down and remember each day.

Always keep your purpose ahead of your goals.

Always look at value from the customer's point of view or from the client's point of view or the department's point of view, not your own.

Back in 1936 a fellow named Dale Carnegie made an interesting comment in his book "How to Win Friends and Influence People" that I try to keep true to this day:

"Always look at everything from the other person's point of view that will keep you out of trouble."

If you do that and you stay purpose-driven, all your goals will be met.

"Anthropologists use this approach when conducting ethnographic research. Their challenge is to leave their own values and mental filters at home so that they can study a particular tribe entirely on its own terms, without prejudice." From "Winning the Long Game" by Steven Krupp and Paul Schoemaker.

Change your thought patterns. Understand what you're trying to do. You want your clients to be successful. If your clients are successful, you will be successful. They will see you as a partner. They will see you as the person they can depend on to do the right thing and help them solve their day-to-day problems.

Goal
You Win

Purpose
Client Wins

Always keep Purpose Ahead of Goal

Chapter 5 - Time for a Re-Compete

Your Company will be attacked.

Ok, you have become a believer in a purpose-driven approach to management. But you are now faced with the inevitable. You have executed an excellent contract and have delivered valuable goods and/or services for a period. Now the customer has decided to open competition with other suppliers. How and what can you do?

Vulnerability

You need to look carefully at the five areas where you will be attacked:

- Performance
- People and Staffing
- Pricing
- Arrogance
- Capture and Proposal Process

Don't skip over any of these. They all present opportunities for vulnerability.

Performance

- Your customer knows all incumbent's flaws and weaknesses but will likely accept competitors "claims of greatness" at face value. How well do you know the customer? Have you pushed things under the rug in hopes that they would just go away? Make sure to address each of the following.

  - Our performance is always more closely and realistically evaluated.
  - Customers don't like being embarrassed.
  - We always believe that we are great — but does the customer agree?
  - Do all levels of the customer organization have a good opinion/impression of Your Company's performance? Who is scoring the proposals? Who is influencing the scorers?

- Unkept promises can kill us! If promises were made in the original proposal but not kept (Have you told yourself, they weren't *really important* to the customer OR we got too busy, and everyone forgot OR the work/scope changed...) realize that customers tend to remember those promises. Broken promises erode trust.

- − Re-promising the same or new things in the recompete won't be believed.

- Don't underestimate the significant role that subcontractors might have in the customer's opinion of your company. Unhappy subs who have customer influence are often saboteurs. Broken promises, not enough work, poor treatment, etc., encourage subs to team with competitors and turn the customer against you.

People and Staffing

- Bidding incumbent staff
  - − You have a great team working on the job, so you decide to automatically rebid the same staff (even if they are just the ones, we're *sure* the customer loves). What is the downside of this approach?
  - − Guarantees a higher price – the competition may bid qualified personnel with fewer years of experience and fewer years of salary increases.

- Sends message of "same old, same old" — The customer may see this as no innovation, no new approaches.
- May put people back on the contract that the customer doesn't like. How sure are you that the customer *actually loves* the staff member?
- Overqualified
  - Are we in the position to chronically overbid people/staffing while competitors will bid to minimum requirements.
- Salary Escalation
  - Are we bound to bid escalations for our incumbent staff salaries?
- Know too Much
  - Do we tend to bid (and price) over the requirements to impress the customer (i.e., we "really know" what they need) rather than following the RFP. Guess what gets scored? Guess what the competitor will price?
- Alternate Staffing Strategies
  - Do we ignore the competitors' potential staffing strategies. Competitors may have a novel approach by changing the labor mix and incorporating more junior staff with a mix of strong senior leadership. Are all your staff what might be considered as "senior?"
- Client Incumbency
  - Have problems when incumbency is longer than tenure of customer personnel. Is there a

new set of decision makers? Are their motivations the same as those of the original customer team?

- – "New" customer might say: Your Company's *experienced* staff is not open to new approaches, can't innovate."

- Proposal Second Team
  - – Proposal teams, under pressure to use available personnel or current people who only meet some of the RFP requirements, gamble that they are "good enough."

- Staffing Quals

  - – You may tend to trivialize the resume/staffing/personnel requirements in RFP — BIG mistake.

- Using Incumbent Staff for Proposal

  - – You rely on incumbent contract management/staff to lead capture and proposal efforts. This presents two problems. First, the staff is not focusing on the customers' current work. Quality may drop off at the most critical evaluation period. Second, existing staff tend to write about the work as they know it, not the work as defined in the RFP. This can result in significant overpricing.

Pricing

Pricing can be affected in multiple dimensions.

- Due to the complex nature of Your Company's business, highly qualified/experienced employees, etc.… competitive pricing is always a challenge. You need to ask yourself, "are our rates comparable to open rates?" Does the customer really want the best when good- enough will do the job?

- It is a fatal mistake to ignore the competitors' potential pricing strategies.

- Remember competitors will try to bid on less expensive people (incumbents are more expensive by default). Smart incumbents know your rates.

- Indecisive pricing leadership, sloppy proposal/capture or process, and poor teaming can all drive price up.

- Existing subs also contribute to price growth. If subs are not adequately controlled, they can quickly drive-up overall price.

    - You are not sure what the customer is willing to pay and what level of service they will accept (vs. language in the RFP). You need to understand unstated bias on perceived value.

- Relying on our "inside knowledge" leads to bidding on what's really needed vs. what's asked for in the RPF − this drives price up. The competitor will not bid to perform work or deliver a service that is not stated in the RFP. It is ok to point out "inside knowledge" but do not price it.

- Confusing performing the work vs. winning the work! RFPs are never 100% representative of what you think to be the "real-world" effort. Proposing and pricing "real-world" usually results in a loss.

Arrogance

Addressing "real-world" in a proposal may come across as arrogance. It can present the impression that you know what the customer needs to buy better than what they are stating that they want to buy. Don't get caught up in this trap. Indicators that you may be subject to the fall include hearing or stating some of the following comments:

- "The customer loves us" (i.e., "they'd never replace us, we know the work better than anyone else").

- The customer is so dependent on our people, they have to re-award to Your Company to get them back.

- We are a trusted partner with our customer (i.e., "they can't do it without us").

- We know more about the work than the customer ("they aren't asking for the right thing in the RFP").

- We're smarter (more technical, more politically astute…) than the customer.

- We don't need to explain how we will perform — the customer already knows.

- We don't need to include a transition plan; the work is already ours.

- Ignore RFP requirements -we know what the customer really wants.

- We can write a better story- don't worry about Section L.

- Responding to the sample tasks isn't really important because we've been successfully doing the work for xx years, and the customer already knows that.

- Ignoring the Gold, Blue, Green, or Red Teams - " they don't understand the RFP, the work, or the customer."

The existence of this attitude will exaggerate the client's perception you are an arrogant contractor, and their life might be better without you.

Capture and Proposal Process

You must follow a disciplined capture and

proposal process. This is easy to remember when pursuing new bids but is often ignored when addressing a recompete. Make sure to address the following potential pitfalls.

- **Poor capture process/lack of customer knowledge**

    - Company-wide: Your Company may be "capture challenged." Unreliable or vague customer, competitive and procurement intelligence, KILLS YOU before you even start. Build and follow the necessary processes to get each in order.

- **Poor teaming decisions**

    - Teamed with wrong company or waited too late (post RFP release).

    - Didn't engage sub early enough in the process (can help with capture, and start past performance, resumes, etc., earlier than Your Company).

    - Keeping subs in the dark during the capture and proposal phases.

    - Not enough input or control of proposal process; not aware of true reputation with customer.

- **Poor proposal process**

- Starting too late: affects teaming, theme continuity, proposal organization/clarity; stresses team.

- Not using or listening to proposal professionals. Skipping or not following the RFP requirements analysis. Not conducting or listening to color reviews (Gold, Blue, Red, Green).

- Proposal infighting or weak or poor proposal management and direction.

- Weak in describing details in transition/staffing plans, tech. approach, sample tasks, etc.

- Wasting space/time/effort on self-congratulatory, historical, and introductory language or unsubstantiated claims of greatness (i.e., Your Company is the industry leader in…).

- **Past performance**
  - Not considering which contracts to reference early; using contract references that don't meet ALL the requirements or aren't easily scoreable as compliant; not verifying customer contact info, account "reference ability"; not touching base with Your Company PM

- **Small business requirements**

- Ignoring or trivializing the RFP's stated small business goals

- **Lack of management commitment**

  – Manager took shortcuts and didn't commit the resources needed for good capture activities, put in second-string proposal team, didn't use available resources for help, and/or didn't ask for BU/Group/Corporate support.

Ok, now that you have identified potential vulnerabilities, what do you do about it? You must address each of the five areas.

Repel the Attackers

We must honestly look at how we have performed for this client in the past. Have we had hiccups? We need to carefully review our performance. Competitors will look for our weaknesses.

It is important that we take the time to understand our weaknesses.

Traditionally we look at ourselves using a SWOT analysis (Strengths, Weaknesses, Opportunities and Threats). You might be better served by changing the order: Weaknesses, Opportunities, Threats and Strengths (WOTS).

WOTS starts with Weakness. Understanding weakness generates Opportunities. Combining those opportunities with an honest assessment of Threat yields Strength.

Exploit your weaknesses and show how you have and are making the changes necessary to alleviate them. It is great to show humility.

Use those corrective actions as examples of lessons learned and value that you bring to the recompete.

I have even volunteered to work with our successor during transition if we are not awarded the recompete. I would share our lessons learned wherever possible. Program success is our number one objective.

Performance

Here are some steps that you need to take.

- **Establish a process to examine performance routinely and objectively.**
  - Develop an action plan to improve based on results.

- Interview both customer management and technical customers.
- Follow up with customer on improvements that were made afterwards.

- **Conduct third-party assessment by outside person/firm 12 months before the RFP.**

  - Provides an independent perspective, verifies internal appraisal results, provides structured means of assessment.

- **Listen to the customer during performance.**

  - <u>Read</u> award fee letters (CPARS, etc.): address negative comments even if award fee % is high.

  - Take critique of our deficiencies seriously and fix them quickly.

- **Show willingness to improve and innovate during contract.**

- **Review original proposal and promises 12+ months before the new RFP is released!**

- **Establish a process to track and implement promises during contract:**

  - Make them part of contract SOW.
  - Accomplish them as an Your Company cost-share investment if you must. Remember, you're investing in the upcoming competition.

- Ensure that customer sees that we are implementing promises and Your Company's investment.

Now that you have identified your vulnerabilities, how are you going to address the staff that has been or is performing under an existing contract?

People and Staffing

- Implement a staff replenishment plan during current performance and coordinate with the customer. Find out which old-timers they want to keep and what new thinking they want.

- Assess every single person's billing to the contract especially the PM and key staff.

- Strongly consider the competition's likely strategy:

  - They typically bid lower priced people OR promise to hire our people post-award.

  - Ensure that the customer believes staff is loyal and tied to Your Company (make sure it's true).

  - Brief staff that competitors may contact them. Competitors frequently "buy in" to the contract; means defectors will likely take a pay cut IF they win (evidence shows -10% or more). More likely, they will just use your experience and inside knowledge to try to win.

- Instruct staff to report all contacts from competitors at once.

- For each Your Company employee we mention by name (especially with resumes) submit an "I authorize no one else to use my name or resume" certification.

• Only submit resumes and staff that meet (not exceed or under-meet) the RFP requirements. Resumes must be well written/formatted and easily scoreable.

• Use incumbent contract management/staff to support, not lead, capture and proposal efforts. They are too close to be objective. They are emotionally attached to work, customers, current staff, etc.

Now that you have cleared up any staffing vulnerabilities, what are you going to do about pricing?

Pricing

• **Pricing must be competitive to win (then perform).**
  - Few customers can justify selecting a higher bidder, even a well-liked incumbent.
• **Get good competitive pricing intelligence early:**

- Your company may have a pricing database with competitor's labor rates (sometimes hard to use, but worth the effort).
- The ARDAK Corporation and its competitors have subscription services that can be used to provide competitor wrap rates.
- **Adopt a straightforward labor pricing strategy/price-to-win strategy.**
  - Price low for labor categories not expected to be filled.
  - Price labor rates vs. real people.
  - Only price the things required in the RFP. Don't let your "inside knowledge" of everything else they need (or the hidden risks) affect your pricing. The competition will be priced only to the RFP's minimum requirements.
  - Be aware that a hungry competitor may "buy in" with a below-cost bid to win. Include the pricing people in the proposal process, especially the strategy and theme meetings.
  - Beware that bidding all incumbent personnel may result in a higher price. Competitors will typically bid the minimum personnel requirements (lower qualifications, lower price)

All right, you have all the pricing issues under control. Now what do you do about team arrogance? "We have performed this job for X years and no one

else can do the job like us." Have you heard that or
similar from your team?

Arrogance

- Ensure strong performance throughout contract,
  as if the recompete were tomorrow.
  - Re-define the opportunity (i.e. budget, mission,
    SOW)
  - Understand where customer is going — new
    mission? New technology? Reorganization?
- Conduct regular and honest customer and self-
  assessments.
  - Request formal performance reviews (e.g.,
    CPARs) and initiate a third-party assessment.
- Keep all promises!
  - Meet with technical leads to see if they made
    promises.
- Make good people and staffing decisions during
  work and when bidding for recompete.
  - Remove staff with customer issues promptly!
- Ensure good pricing on current work and when
  biding for recompete.
  - Develop a strategy for reducing the number of
    staff with high labor rates.
- Keep up with political contract/mission changes.

- Guard against competitors.

  - FOIA our own information to see what competitors can/will learn about us.

  - Find Your Company experts to keep yourself alert, on track, honest, objective — and listen to them.

Now that you have stopped drinking your own bath water, it is time to roll up your sleeves and go to work. You are ready to start your Capture and Proposal Process.

Capture and Proposal Process

Here are the minimal activities that you must undertake.

- Develop a dynamic capture plan — early!
- Vet your win strategy with a win strategy review — be prepared to revise your strategy.
- Get management buy-in and commit dedicated/experienced proposal resources (i.e., capture manager, technical lead, proposal manager, book bosses/volume leads, etc.).
- Hold capture status meetings.
- Create a capture schedule to ensure accountability and capture/proposal deliverables.
- Conduct a Black Hat workshop/competitive analysis.

- Price-to-win early (e.g., independent cost estimate, BOEs, BOM, risk across est. elements).
- Conduct a Technical Readiness Review (TRR) to focus on technical solutions.
- Conduct a Proposal Readiness Review (PRR).
- Identify and select subs by first understanding the company's capabilities and knowing the customer's favorite contractor/vendor — early!
- Include your teammates in the capture/proposal process.
- Use the knowledge of our contracts, subcontracts, and pricing folks. Invite them to the table.

Don't Know Too Much

Now that you are ready to put pen to paper, it is easy to know too much. If you are responding to a request for proposal, remember, the competitors will bid against the RFP, not reality. You may be aware of factors that are not included in the solicitation and want to address them in your response, DON'T. Remember:

- An RFP is not reality!

- It is a model that captures the essence of the program.

- Competitors bid against this limited model, not reality.
- If we bid against reality, we will not win in today's cost-sensitive environment.
- It is ok to illustrate our understanding, just don't price it.
- When we win, life can go back to normal.

Now that you are ready to go after the recompete, take the time for the following steps.

- Catalogue our real and perceived weaknesses.

- Take the opportunity to address and where possible correct the weaknesses before the RFP/Proposal.

- Be paranoid - look for threats and report them to the capture manager.

- Start building stories about how we have identified and addressed weaknesses over the life of the program (write them down.).

- Collect any citations and metrics (Don't worry about significance, just collect and catalogue).

- Identify all stakeholders - make the time to personally contact them and ask what we can do better.

- Win!

There is significant work to be done. Who is responsible for leading the charge? I have found that a dedicated capture manager should be assigned to any complex, must-win bids (both recompetes and new opportunities).

Once you have established your purpose and found potential clients that align with the values that you bring to the table, you need to assign a capture manager to bring the opportunities home and generate revenue.

I recommend that you conduct a group session and address each of these issues as you attempt to qualify an opportunity.

I provide a PowerPoint Qualification template on my website: *ResponseResource.com under tools.*

Remember that the other man may be totally wrong. But he doesn't think so. Don't condemn him. Any fool can do that. Try to understand him. Only wise, tolerant, exceptional men even try to do that.

There is a reason why the other man thinks and acts as he does. Ferret out that hidden reason – and you have the key to his actions.

Try honestly to put yourself in his place.

Dale Carnegie

Chapter 6 – Client Engagement Process

"What We Have Here is a Failure to Communicate"

Cool Hand Luke

In 2015, I developed a PMI certified eight-hour training course discussing client engagement. The following discussion has been excerpted from that course. The course summary can be found in appendix B.

Client engagement helps us to better understand the why of a system before we can implement using a purpose-driven approach to management. This presents a problem. We must work with the client and the various levels of stakeholders. Stakeholders often fill the role as sponsors.

In a 2021 Forbes Business Development Council article, Majeed Hosselney reminds us:

"For example, I'd underline the importance of sponsorship as the air a project breathes. To build the road ahead, a project leader generally depends on proper sponsorship by business leaders. To be successful, a project also needs to achieve a balance between focusing on not only technical solutions but also the people who are the target for the solution and the processes supporting the solution's growth."

If we are left-brain dominant, reaching out to "people" is not our favorite pastime. We would prefer to read a set of requirements and specs and build.

Jay Grusin in *"Intelligent Analysis"* points out another problem in pursuit of "why".

> *"We find developing key intelligence questions challenging because we assume that we know the client's requirements and we know what they want to hear."*

In a recent article, James Clear discussed additional common management problems:

1) Your Client Gives You Vague, Ever-changing Requirements.
2) Your Client is Slow with Communication.
3) The Project Doesn't Start on Time.
4) You Try to Manage Every Project the Same Way.
5) The Client Doesn't Like What You Created.
6) Your Point of Contact Doesn't Seem to Care About Your Project.
7) Too Much Time is Spent Solving Problems After Projects Are "Live".
8) Your Company Wants to Grow the Business and Asks for Help in Finding New Opportunities.

There is a common theme that permeates these major (and common) challenges. We are not talking with the right people (stakeholder management), and we are not communicating effectively. Chapters 10, 13 and Appendix X3 of the Program Management body of knowledge address these issues.

Steve Yager, CEO Artemis International, further highlights these two underlying areas of concern:

Stakeholder Management.

Effective stakeholder management requires the identification of individuals who are affected by and/or can affect the successful outcome of a project, especially those who are of a less than positive disposition toward the project objectives. All stakeholders require attentive management to minimize obstacles of this type.

Solution: Create a truly collaborative work environment. Visibility into the work involved is likely to result in change. Collaboration allows the project to be analyzed and discussed by all interested and affected parties. This will ensure minimal uncertainty and provide the wherewithal to keep all interested parties "on board". Ownership of risk identification, planning, management, and tracking is paramount. This information must be published and provided to the appropriate stakeholders.

Communication breakdowns cause unclear project goals and objectives.

Management may rethink its goals for a project, not communicate them well and expect the team to adapt accordingly.

Solution: Working without an up-to-date, well-stated purpose can blur project focus and demotivate the group. Having a well-crafted purpose statement helps to highlight, record and track enterprise-level and project-level objectives and communicate them in an understandable manner. Always begin with documented criteria for measuring success. Require the project sponsor to define a measurable result. Not only will this increase the chances of project success, it also will aid in project scope management. To avoid communication breakdowns, project managers also should facilitate good communication — conflict resolution, coordination, and empowerment.

Whether you are pursuing a recompete or preparing to bid on a new opportunity, you must shape and track the relationships with the client.

Patrick Lencioni outlines "The Five Dysfunctions of a Team" in his book by the same name. His pillars of dysfunction include:

Dysfunction	Results
Absence of Trust	Invulnerability
Fear of Conflict	Artificial Harmony
Lack of Commitment	Ambiguity
Avoidance of Accountability	Low Standards
Inattention to Results	Status and Ego

These dysfunctions will impact the ability of the team to engage with the client. As you work with your client, you must be aware of each. Start by building trust with the client, putting your fear of conflict aside, making firm commitments, setting high performance standards, and ensuring that you follow through with valued results.

Establish trust in client relationships early in project development. Shape and track those relationships over the life of the project and beyond.

Projects may come and go, but clients will likely be around longer. If you manage the project with purpose, trust is a residual benefit.

You need to shape and maintain client relationships.

Shaping and Tracking Relationships

You must identify stakeholders and reach out to them. They are unlikely to reach out to you.

- You must ethically shape relationships to the benefit of the client.

- You must include the shaping process in your collection plans.

- You must identify stakeholders with the authority to request requirements.

Client engagement can be broken down into four phases:

1. Doing your homework,
2. Preparing for the engagement,
3. Conducting the engagement, and
4. Documenting the engagement

We often find ourselves skipping steps 1,3 and 4, jumping right into a meeting.

Do your homework!

In addition to identifying stakeholders, you must understand their role. Learn before you go. You need to anticipate their needs but be careful not to project. Your preconceived ideas may not be shared. Moreover, they may be wrong.

"There is nothing more frightful than ignorance in action."

Johann Wolfgang von Goeth

Prepare for the engagement.

Take control and actively set appointments, emphasizing the importance of collecting the stakeholder's position.

Before the engagement, clearly understand what you know and don't know. Build a list of questions that you want to discuss. I have even sent the questions ahead to give the stakeholder time to consider them.

Engage

Start by establishing a trusted relationship. Negotiate rules, rights, and responsibilities of the relationship. Talk about on and off record findings. Build a sense of mutual respect.

Ask for permission to ask questions. Ask permission to ask about issues and pain. Establish a working relationship. Ensure a sense of mutual respect and openness.

Work to qualify or disqualify the requirement. If the requirement appears to fall within the scope of the contract that is great. If not, work with the client to formulate potential alternative solutions.

As discussed earlier, contracts will likely get into the act if the new requirement appears to be out of scope.

Share your purpose and goals. Ensure that you communicate that your purpose is to see their world from their perspective to determine whether the project will address their needs and create value. Your goal is to determine whether you can incorporate a solution to their needs withing the scope of the project.

Remember that active listening is difficult. Learn and practice your interpersonal dialogue skills. Those skills include Socratic questioning, reversing, and nurturing.

Socratic Questioning

Asking a series of questions, whereby the prospect discovers the desired knowledge by answering the questions.

Reversing

Answering a question of statement with a question.

Nurturing

Questioning and reversing without nurturing is interrogating. Help the interviewee to stay or feel ok.

Thomas A Harris, M.C. published a book, "I'm OK – You're OK", that discusses how to make someone feel ok.

I like to use what is called intelligently asking dumb questions. It makes the interviewee feel ok.

There is a difference between acting dumb and being dumb. There is a limit to how smart you can be. There isn't a limit to how dumb you can be. Don't look too good or talk too wise.

You will be perceived, positioned, and treated at the level of your thinking. Credibility and trust are established by the questions you ask, not the statements you make.

If you start the process well, the process gets easier. If you start poorly, the process gets harder. Practice makes perfect.

Keys to Success:

- *Ask questions – don't assume anything.*
- *Listen more than you speak.*
- *Thank the interviewee for their time.*
- *Confirm the next steps.*

Failure to follow-up is an opportunity missed to understand needs from a stakeholder's perspective. Remember, people love to be interviewed.

Where do we fail in the elicitation process? **We don't listen well.**

- Under pressure we default to speaking.
- We think faster than the prospect speaks.
- We tend to think about our response.
- It's challenging to listen, take notes and formulate the next question.

No thinking! Thinking comes right before trouble - listen. You can improve your skills by practicing in low-risk situations, taking notes, and focusing on the client, staying focused, following up on the present question, paraphrasing to review what you are hearing.

Thinking

"The thinking that got us here, is not the thinking that is going to get us where we need to be."

Albert Einstein

Don' listen with your motor running!

Document the engagement.

Documenting the results of an engagement is essential. Make sure to capture key details, the client's perceptions, and their needs. Remember your purpose - providing value to the client.

Evaluate Results

The hardest task is to collect all that you have learned and determine its impact on client value. You must evaluate the results of the engagement without bias, without preconceived notions. Ask yourself, "what have I learned?"

Have you identified both hard and soft requirements? Have I identified requirements that will cause implementation problems?

As you look at the results of your effort, you must also evaluate your own performance. What could you have done better? What will you do differently next time?

Document your lessons learned and share them with others so that the next capture will be better.

Remember, you will be perceived, positioned, and treated at the level of your thinking. Credibility and trust are established by the questions you ask, not the statements that you make.

Chapter 7 - Capture Managers

What is a Capture Manager and what role do they play in winning? Complex must-win bids require dedicated attention to win.

A capture manager is a dedicated person assigned to winning a business pursuit, by providing the leadership, coordination, resource allocation, strategy, and management needed to prepare the winning bid. While a business development manager is responsible for a portfolio of leads, a capture manager is usually assigned to just one. Capture Managers are most often used for large complex sales, where closing the sale will take the business development manager's attention away from pursuing the other leads. _**A Capture Manager plays a key role between business development and proposal writing**_ that can greatly improve your chances of closing the sale.

Hardest Job

Being a capture manager is the hardest job in the entire company. It also requires the qualifications and experience that cover a wider range than most people have. Consider:

> *Being a capture manager is the*
> *hardest job in the entire company.*

- A capture manager must know what to offer, how the offering should be designed, and how to make cost/value tradeoffs and still win.

- A capture manager must not only have the technical skills to know how to do the work, but also must know how to sell. A capture manager must not only know how to sell but also how to do it externally as well as internally. A big part of a capture manager's job is negotiating and persuading people within their own company.

- A capture manager must know pricing and contracts well enough to negotiate terms and win.

- A capture manager must know enough about proposals to anticipate the information required to write the winning proposal.

- A capture manager must have enough authority to make decisions and work through other people without being constantly second guessed.

- A capture manager must be able to take risks.

On top of all that, a capture manager has a high chance of getting fired if the bid loses.

Carl Dickson, the Founder and President of CapturePlanning.com and PropLIBRARY provides some excellent additional pointers:

- Marketing vs. Business Development vs. Sales vs. Capture Management
- What to focus on first when considering a capture process
- Are you missing the two most important ingredients for transitioning from sales, business development and capture to proposal writing?
- Business development vs. capture management vs. proposal management vs. winning
- Capture manager
- 8 pursuit and capture process goals to accomplish before your proposal starts
- What is a Capture Manager and what role do they play in winning?
- 37 problems to solve for successful pursuit and capture
- Staffing your pursuit effort
- Why the job of capture manager is impossible and how your fate is determined by what you do about it?
- What matters more: business development, capture management, or proposal management?
- The truth about customer intimacy
- How to steal contracts away from the current incumbent

- Find out how selling in writing is different than selling in person
- Whose job should it be to win?
- How to turn your customer, opportunity, and competitive intelligence into winning insights
- How implementing capture management can change your entire company
- Is it business development's job to win?

Assigning a capture manager who can't check all those boxes lowers your chances of winning. Companies that automatically assign the future project manager as the capture manager will usually be assigning someone with skill gaps. If it's a recompete, the project manager may also have bias regarding the current state that prevents them from being sufficiently innovative.

The difficulty in finding a top-notch capture manager is one reason companies often turn to consultants. But this isn't a perfect solution either. A capture manager needs to know your company's history, resources, and capabilities. In some companies, that information can't be quickly learned in just a few meetings. But for some consultants, it's possible.

A capture manager also needs to know the client's preferences. Some companies don't even know their own preferences. Generally, you will not get that by hiring a consultant, even one with some experience at the client's organization. You might get it, however, if the capture manager can spend enough time getting to know the customer. Unfortunately, consultants usually are not given the months that are needed for that.

Most companies go for what is cheap and convenient, and then convince themselves it's an effective choice. Don't be that kind of company. Be the kind that understands win rate mathematics and how achieving a high win rate is about return on investment. Manage to get the highest return instead of the lowest investment.

But mostly, being a capture manager is about negotiations. He or she must collaborate with the internal company organizations to decide where to put the resources and efforts that are necessary to bring home the win. Oh, and then there is the customer.

The customer is easy. All you need to do there is to discover their preferences and give them what they need in a way that is better than their alternatives. It's your own company that will be hard. Who is going to discover the customer's preferences?

- How much effort will be required to do things and where will the resources come from?

- Who has the authority to decide every little thing?

- How do you get the price low enough to win? What is a price that is low enough to win?

- How do you get it all into your proposal, articulated effectively, and submitted on time?

- Oh, and if you think the customer is as easy as it says above, think twice.

All opportunity in your company comes from growth. This means all opportunity comes from capture. It's a hard job. And everybody depends on it. It's worth a quality approach. Don't skimp on the qualifications of your capture manager. Make the person you need available. But don't just assume that getting the right person is all it takes. Hiring the right people is not enough. Surround your capture manager with all the organization, process, and resource advantages that maximize your win rate. Put the amount into it that is necessary to be successful.

If you have the right leader with the right skill set, the capture manager needs to understand their primary responsibilities and be empowered to carry them out to fruition.

Responsibilities

The Primary Responsibilities of a Capture Manager:

- Communicate
  - Establish and maintain contact with BD and the client where possible.
  - Keep management and team up to date (Good news and unwelcome news)
  - Ensure that conflicts are resolved.
  - Facilitate team meetings.
- Take on the role of surrogate customer and ensure that we satisfy the client's need!
  - Understand the client's mission.
  - Understand the client's requirements.
  - Understand the client's constraints.
  - Understand underlying pain.
- Act as the proposal's solution architect (understand what it takes to win the job. One approach is to diagram the solution (Fishbone Diagram Attached).

The capture manager and the team need to look at the opportunity to ensure that it is qualified. It is not enough to say, well it is out there, we can bid. We must have the ability to win.

Resources are constrained, both financial and human capital. We need to discipline ourselves to limit pursuits to those that create value for our clients.

We need to look at opportunities from multiple viewpoints:

- Technical
- Organizational
- People
- Financial
- Past Performance

Can we provide value to our client? Are we able to win?

It is also helpful to use these same factors when determining how well our competitors stack up. We need to ask ourselves some tough questions and provide honest answers.

Technical

- Do we understand the customer's performance statement?
- Do we have the subject matter expertise to do the job and create value for the client?
- Do we have the existing core capabilities?
- Does this opportunity map to our organization strategic plan?
- Have we validated our approach with the client?

Organizational

- Is this a current client?
- Have we had successful engagements with the client?
- Do we understand the client's decision process?
- Are we a viable contractor to get the job done?

People

- Do our people know the client?
- Have we developed a level of trust with this client?
- Do we understand underlying client pain points?
- Have we been dealing with a decision maker?

Financial

- Is the opportunity funded?
- Is this a lowest cost technically acceptable bid?
- Can we meet our profitability goals?
- Are we able to meet the investment constraints?

Past Performance

- Have you delivered to this contract?
- Have you done similar activities for others?
- Have you provided referenceable value to others?

Have we been honest? Can we respond in the affirmative to each of those questions? (Probably not!)

Don't dismay. If the answers are not in the affirmative, can we do something about it? Do we need to conduct better client outreach campaigns? Do we need to add teammates to shore up the team?

It is ok to say no to bad opportunities. I cannot count the number of losing bids that were pursued because someone said, "I think we can win. What have we got to lose?"

Ok, you have made Up Your Mind, Craft the story.

Synthesize the big picture from the multiple team viewpoints.

Extract the major features and benefits that your team affords your client.

Ensure that the team understands a clearly defined set of themes/messages that we need to convey to the client.

- **Orchestrate resources.**
  - Locate mission SMEs.
  - Locate technology SMEs as required.
  - Ensure that all gaps are filled with the best qualified resources including teaming where appropriate.
  - Make sure that contracts and subcontracts are onboard early and kept informed.

- **Act as product manager for the proposal to ensure that a quality product is delivered.**
  - Make sure that the proposal conveys the story, themes, and value. Value is best stated in clear value propositions. (see next section)

One tool that works well in visualizing a solution is a fishbone diagram. Kaoru Ishikawa was the originator of this technique, but the name fishbone replaced Ishikawa Diagram.

An example is provided on the next page.

I see the role of capture manager as one of facilitation, ensuring that all the above is completed and where appropriate taking the action in hand and executing whatever work is required.

A major role of the capture manager is to ensure that the offering conveys a defendable Value Proposition.

Capture Manager
- **Opportunity Manager**
- **Cheer Leader**
- **Plumber**

I have used Ishikawa (fishbone) diagrams as an aide to pictorially display why my organization is well suited to do the work.

Try drawing diagonals that represent the critical areas. Then show branches and limbs off those diagonals that represent experience and proof that the team is ready to win.

Does This Opportunity Qualify?

Use the diagram to prove to yourself and others that you have the ability to WIN. You have the people, processes, technology, and past performance necessary to pursue this opportunity.

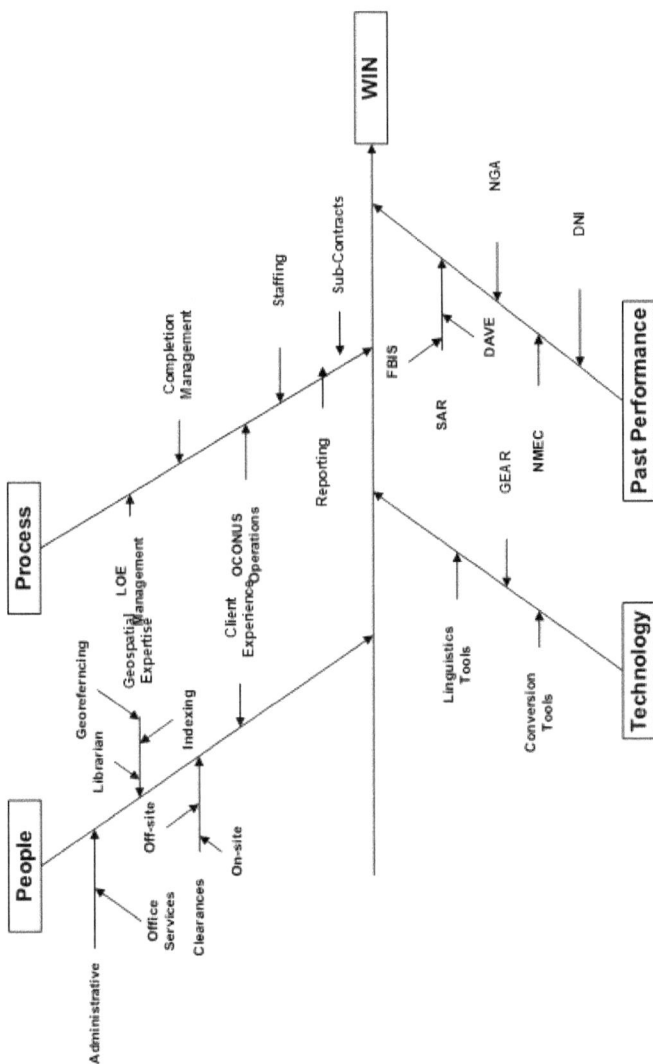

Sample Ishikawa (fishbone) Diagram

After proving that you have the capability to do the work and provide tangible value, you need to look at the value from the client's perspective. You need to develop and understand your value proposition. Remember, value is from the client's point of view, not the supplier's.

We often skip this step and go straight into "this is obviously the best solution". At least it is obvious from our point of view.

I once had a student that wanted to understand how to work with a somewhat difficult client. She was sure that her solution would bring great value to the client. The client only needed to put her under contract and let her deliver. If he did so, he would eventually see the great value that she forecast.

I tried to explain that clients usually must see the value before they buy.

Always do right. This will gratify some people and astonish the rest.

Mark Twain

Chapter 8 - What is Value?

Value is in the eyes of the purchaser, not the seller. We often forget that and try to sell what we have, not what the customer needs.

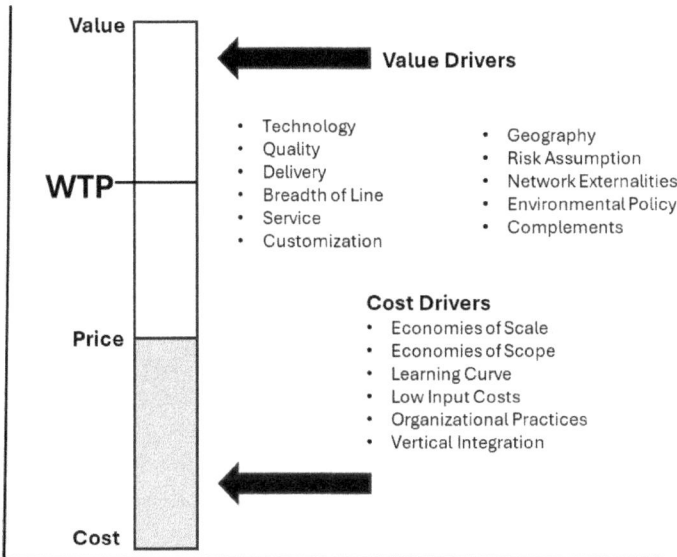

WTP – Willingness to Pay (what is it worth?)

How does a customer determine Value?

- equivalent worth or return in money, material, services, etc.: to give value for value received.

- the worth of something in terms of the number of other things for which it can be exchanged or in terms of some medium of exchange.

- the desirability of a thing, often in respect to some property such as usefulness or exchangeability; worth, merit, or importance.

- Absence of pain in making a buying decision.

Creating Value Propositions

What are the drivers for value, and for that matter, price, and cost? Also, how much is the customer willing to pay?

What factors act on customer's perception of value. Michael E. Porter in his book, "Competitive Advantage) provides the following chart that outlines those factors.

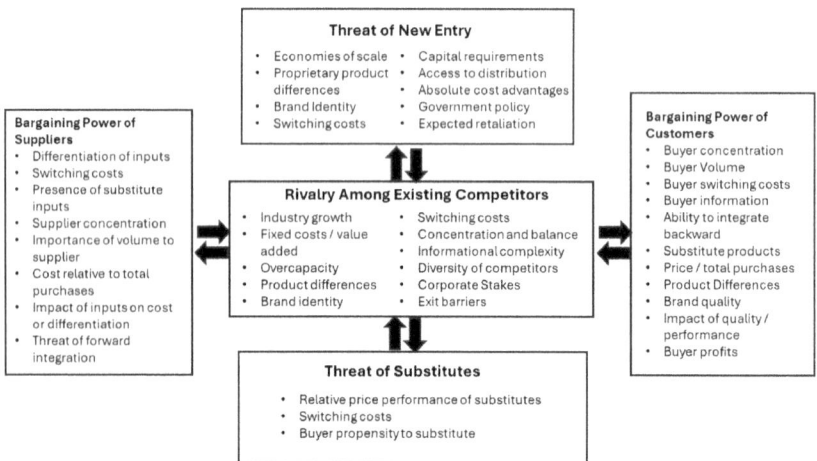

Threat of New Entry

- Economies of scale
- Proprietary product differences
- Brand Identity
- Switching costs
- Capital requirements
- Access to distribution
- Absolute cost advantages
- Government policy
- Expected retaliation

Bargaining Power of Suppliers

- Differentiation of inputs
- Switching costs
- Presence of substitute inputs
- Supplier concentration
- Importance of volume to supplier
- Cost relative to total purchases
- Impact of inputs on cost or differentiation
- Threat of forward integration

Rivalry Among Existing Competitors

- Industry growth
- Fixed costs / value added
- Overcapacity
- Product differences
- Brand identity
- Switching costs
- Concentration and balance
- Informational complexity
- Diversity of competitors
- Corporate Stakes
- Exit barriers

Bargaining Power of Customers

- Buyer concentration
- Buyer Volume
- Buyer switching costs
- Buyer information
- Ability to integrate backward
- Substitute products
- Price / total purchases
- Product Differences
- Brand quality
- Impact of quality / performance
- Buyer profits

Threat of Substitutes

- Relative price performance of substitutes
- Switching costs
- Buyer propensity to substitute

Porters Five Force in "*Competitive Analysis*", 1985

In a competitive environment, there are many forces in play that also impact our ability to communicate value.

1. Competition
2. Barriers to Entry
3. Buyer Power
4. Supplier Power
5. Substitutes

Robert Cortucci discusses how these forces affect competitive intelligence in his book by the same name. Each of these factors drive the value/price comparison (**VPC**). We must remember to address each of them.

1. How does intense competition affect VPC?

More intense competitive rivalry...

- May reduce price, while holding value and cost constant

- May increase value (+cost) without increasing product prices.

How do we reduce competitive rivalry?

- Cutting costs (in ways customers don't notice) may stop rivals from trying to compete on price (retaining higher profitability)

- Compete on non-price dimensions.

- Differentiating products reduces competitive pressures on price.

- How easy is it to imitate our key to differentiation; how good is our defense?

- Even if replicable, ongoing innovation in value or cost drivers can reduce competitor pool if missteps lead them to exit.

- Create switching costs (that customers don't notice or mind)

- Focus on niches, market segments with competitive advantage.

2. How do low barriers to entry affect VPC?

Low barriers to entry...

- May reduce prices while holding value and cost constant.

- May increase costs by bidding up resources (pass-thru to prices)

How do we raise barriers to entry?

- Decrease costs and prices through investments in scale.

- Design switching costs into our offering.

- Increase barriers to imitation.

- Establish property rights, acquire dedicated assets, increase causal ambiguity, increase learning and development costs.

- Signal ferocious rivalry through credible threats.

  - Willingness to take losses for a long time (deeper pockets)

  - Huge investments in capacity (burn the ships)

3. How does buyer power affect VPC?

Powerful buyers...

- Increase the value (and cost) required to compete.

- Reduce the price required to compete.

How can we reduce buyer power?

- Avoid falling into the trap of serving a small number of buyers.

- Differentiate product from competitors.

- Reduce dependence by serving multiple market segments.

- Help buyer structure value chain around product's unique features.

- Branded demand stimulation with end customers

- Reduce buyer's threat of backward integration.

- Consider forward integration into buyer's industry.

- Avoid investments in capacity expansion (unless they can lower prices and decrease buyer power by driving out competitors)

4. How does supplier power affect VPC?

Powerful suppliers...

- Increase costs.

- May decrease the value a firm can offer to customers.

How can we reduce supplier power?

- Aggregate orders into very large, infrequent purchases

- Innovate to develop substitute production processes that do not rely on traditional suppliers.

- Develop credible threat of backward integration (or actually do it)

- Avoid branding efforts by supplier to end customers.

- Resist relying on differentiated features of supplier's products.

5. How do substitutes affect VPC?

Powerful, viable substitutes…

– Increase value required to compete.

– Lower price required to compete.

How do we reduce the threat of substitutes?

– Find ways to boost value by less than cost of doing so, or lower cost by more than the associated impact on value.

– Raise buyer's switching costs for your product (unique features, brand loyalty) or for your industry (branding your product class)

– Be the first to create substitutes (shoot your own dog)

– Acquire patents developed by potential substitutes.

– Organize a collective industry response (we're all in the same boat)

As you look at each of these factors, ask yourself the following questions. Or even better, pull a team together to work on them.

- **How can we reduce competitive rivalry?**
- **How do we raise barriers to entry?**
- **How can we reduce buyer power?**
- **How can we reduce supplier power?**
- **How do we reduce the threat of substitutes?**
- **What are our value propositions?**

I try to use a simple approach to determine the desirability of a thing from the customer's perspective. I took this from Dale Carnegie's Golden Book:

> *"Rule 17: Try honestly to see things from the other person's point of view."*

What does the customer want and need, not what do we want to sell. If we are to do that honest assessment, there are some questions that need to be answered:

- **Who is the customer?**
- **Does the customer know us?**
- **What does the customer state are his wants?**
- **What does the customer infer he wants? (What does the customer value?)**
- **How are we going to give the customer what he wants?**
- **How can we prove that we can deliver?**
- **How is the customer going to benefit?**

The customer needs to understand, what is in it for me? (WIIFM?)

As we build our value proposition, we need three strong ingredients:

- We need to Resonate.
- We need to Differentiate.

- We need to Substantiate.

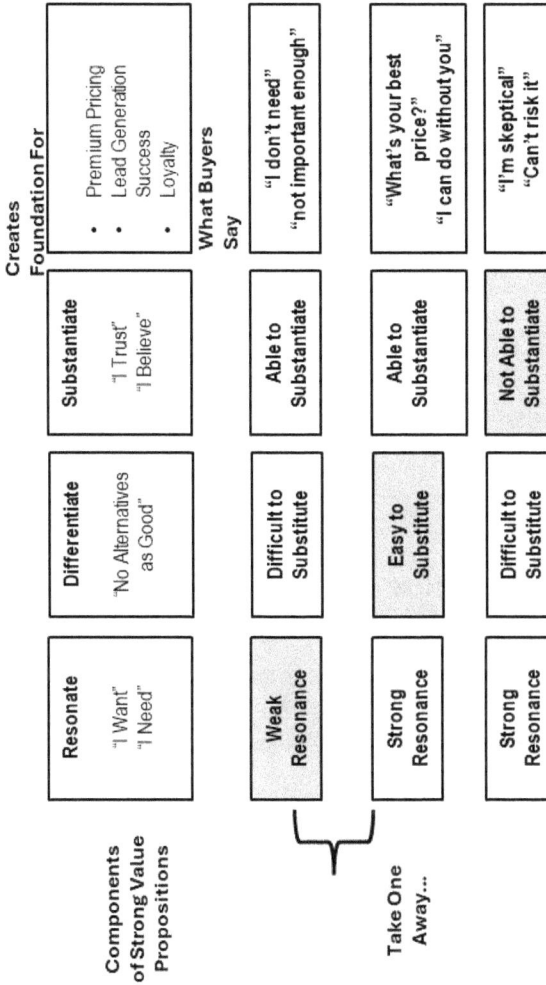

Components of Strong Value Propositions	Resonate "I Want" "I Need"	Differentiate "No Alternatives as Good"	Substantiate "I Trust" "I Believe"	Creates Foundation For • Premium Pricing • Lead Generation Success • Loyalty	What Buyers Say
Take One Away...	Weak Resonance	Difficult to Substitute	Able to Substantiate		"I don't need" "not important enough"
	Strong Resonance	Easy to Substitute	Able to Substantiate		"What's your best price?" "I can do without you"
	Strong Resonance	Difficult to Substitute	Not Able to Substantiate		"I'm skeptical" "Can't risk it"

We need to understand how the customer views those components and how each is important.

Here is a simple to follow Value Proposition Template:

- Customer will "get" - "realize"-"enjoy"
  - *Specific benefit.*
- Because our approach, product, service provides.
  - *Added value statement.*
- We have the ability to provide this value because
  - *Proof of our ability.*
- Where have we done it before
  - *Example of prior work.*
- Conclusion
  - *Summary of value proposition.*

Example filled in template:

"Customer" will enjoy

> *an expanded breadth and depth in training for ever evolving tools and techniques as technology progresses.*

Because our approach brings added value of

> *combining our strengths in system engineering, system integration, instructional design, and training.*

We have the ability to provide this value because

> *we have built a team that has experience solving similar problems.*

We have proven this approach is effective

> *by the work we have performed in developing and delivering training for CASE1, CASE2 and CASE3.*

Conclusion

By coupling our instructional expertise with the underlining disciplines of planning, engineering, and integration, we are able to provide the best available training for our "customer".

Appendix C outlines a slide deck that I have used in team exercises to develop winning value propositions. Feel free to download and use as you see fit. (no charge!)

With your value propositions in hand, how does your competition compare?

Multi-Attribute Competitive Analysis

You have determined that you can provide value to your client and your organization has decided to pursue the project. You need to determine if others might also compete for the same project.

Are you competitive?

Remember, your competitive position is directly correlated with the client's perception of the value that you or a competitor might offer.

An objective approach in determining how you stack up against your competitors is to use multi-attribute utility analysis (MAU).

Your first step is to determine the factors that your client might use in evaluating your offer vs your competitor's. If you have a request for offer in hand, that would be a good starting point.

List the factors that the client has indicated will be evaluated. The second step is a little more creative. Assign weighting to each of the factors. If you are lucky, the client has provided some guidance on the relative importance of the factors.

I have found that this is a good group activity. Walk through each factor and ask who the best is. Who is the worst? Give the best a score of five and give the others a relative score. You may have more than one five, and all others are threes or fours.

Multiply the weight of the factor times the score for each competitor. Don't forget yourself.

Sum the compute scores for each competitor and divide by five. You now have a normalized score for each competitor. The higher the score, the more likely that they will become the successful bidder.

Your scoring matrix might resemble the following example. I have included a section M from a government solicitation to define the factors. An abbreviated Section M (sample evaluation criteria) can be found in appendix A).

	Factor	Weight
F1	Technical Understanding, Capability, and Approach	30%
F2	Management Capability	25%
F3	Staffing and Personnel Qualifications	10%
F4	Past Performance	15%
F5	Cost/Price	20%

Factor	Co. A	Co. B	Co. C	Co. D
F1	5	3	2	4
F2	4	5	2	2
F3	3	2	5	1
F4	2	3	3	5
F5	5	4	3	3

And the Winner Is

Company A	**3.10**	Company C	1.85
Company B	2.80	Company D	2.55

Once you have a cut at the utility model and determine that you are losing competitive points in one or more factors, you may want to think of adding a teammate.

One thing to keep in mind - if the scores are the same across competitors, the factor is no longer a discriminator regardless of the weight applied.

I first used MAU in 1975. I was consulting for the Army's Tank and Automotive Command evaluating proposals for a new weapon system. We had broken the factors as defined in section M of the request for proposal, assigned weights, and scored the three competitors. One evaluation factor had a high "weight" and one of the competitors received an exceptionally low score on that factor because the component failed in prototype testing.

One colonel complained. He said he had been using a similar component from the same contractor for twenty years and the component never failed. We changed the factor score to effectively nullify the impact of that criteria. The factor was no longer a discriminator.

Unfortunately, it did not help the competitor; they lost.

Remember Teaming Can Build Strength

Find teammates that you would consider to be better than you as it relates to a specific factor. Add them to the team and calculate the resulting score.

If there are many factors, you may need to add many teammates to put your team in a competitive position.

As your team is conducting the sample scoring, use the client's definitions where available. If they provide a discussion of evaluation factors, use theirs, not yours.

If you want to take the uncomplicated way out, you may download an excel tool rather than writing your own. You can find a MAU model on my website at:

ResponseResource.com (select the tools option)

You are likely to have many more factors and possibly more than four competing companies. This process helps you to understand where you fit in the competitive arena.

There are several excellent books and articles on Multi-Attribute Utility Analysis. If you would like to dig deeper, you will find several candidates in the reference section of this book.

Check out "Competitive Intelligence", Robert Cortucci.

Who is not satisfied with himself will grow, who is not sure of his own correctness will learn many things.

Chinese Proverb

Chapter 9 - Congratulations, You Won!

Scope Exceeds Contract Value (Now What?)

The team has done a great job and congratulations, you won. Oh no, you won. You are sixty days into the engagement, and you find that the true scope exceeds the contract value.

Antonio Nieto-Rodriquez makes the following observation in his article in Harvard Business Review: "Does your project have a purpose?"

> *"All project management methodologies demand that projects have a well-defined business case. But when evaluating and prioritizing projects, looking at the business case alone is not enough. We also need to understand how the project connects to a higher purpose. Evaluating on purpose can help leaders decide whether the project aligns with the organization's strategic goals. It is also a key driver for engaging team members and the organization as a whole and motivating them to support the project. Companies must learn how to articulate a project's purpose. A straightforward method of uncovering it is to simply ask, "Why are we doing the project?" Then, when you've arrived at your answer, ask why again. With each successive layer of why, you will come closer to the project's purpose. If this*

exercise does not help you discover something that will motivate people to work on the project, you probably should not start it."

Antonio Nieto-Rodriquez
Harvard Business Review

The client was sure that they had a well-defined business case and just expected us to read their minds and build the system.

Oh No, We Won!

Different people see things from multiple points of view.

- My company saw a great new contract worth $212,000,000.

- The Contracting Officer sees a successful competition with an award to a quality company.

- The stakeholders saw their view of the world as different from the contracted requirements.

- The program manager and I saw a hard problem to solve.

I was faced with this dilemma. I was Chief Engineer on the program and was responsible for building the solution. I had eight department heads and 32 control account managers. We were following an earned value management system, ready to get to work.

The contract was originally awarded for $212

million. That sounded fairly good to my company. The program manager felt that the requirements were somewhat ill-defined but felt we could do the job. I had worked with the PM before, and he asked me to join the program as Chief Engineer and lead the technical side of the program.

I jumped on board and rolled up my sleeves. The PM agreed to manage the business side and manage subcontractor relationships. We had a nightmare team of 41 subcontractors. I was happy to not have to deal with the subcontract issues. That job was off of my plate.

We agreed that I would treat the entire staff of 350 engineers and developers as a badge-less group. I even required that every member of the team wear a badge with the project logo and not the logo or name of their parent company.

So far, so good. The client was happy. They had worked through the procurement process and now their problem was ours. They had received a reasonable bid from a reputable, proven company.

We needed to determine whether we were going to be goal oriented or purpose driven.

We later found that they had planned on spending $500 million on the project. That was an early indicator that something was wrong with the bid. We should have known that we were in troubled waters. The customer was happy to think that they were going to be able to resolve their problems with a $212 million solution.

How did we get into this mess?

The underlying problem was that different people have different opinions about requirements. Different stakeholders have differing points of view. As we move further into the Agile environment, this problem becomes more obvious.

Solving this problem requires that the program and project managers take the time to understand the "why" of the project and understand the value that is to be realized by the client.

A lot of meaning was lost by the time that requirements were collected at the stakeholder level and passed on up through contracts and up through the procurement cycle.

LinkedIn has created an AI driven blog that is attempting to answer the question:

How do you define the scope of a project in a clear and measurable way?

The first step that they recommend is to understand the project vision.

Understand the project vision!

We asked ourselves, "Why are we building this system?"

Before you start defining the scope of a project, you need to understand the project vision. This is the high-level purpose and objective of the project, and it should answer the questions: Why are you doing this project? What value will it deliver to your organization or customers? Who are the key stakeholders and beneficiaries of the project? The project vision should be concise, specific, and aligned with your strategic goals.

We found that what we were being asked to do under the contract was not what was being originally asked for by the stakeholders. Have you ever run into this problem?

The challenge at hand is "how do we bring this mess under control?"

How did you bring the mess under control?

- Work to understand the "Why" of the project.
- Review and validate the Basis of Estimate and reallocate budgets.
- Understand where we are in the Project Life cycle?
- Choose a project management style.
- Review and validate the requirements.
- Keep the contract-shop informed.

We needed to better understand the benefits customers were going to receive after we delivered. $212 million is a lot of money. It's a lot of people. There are a lot of activities.

And as I said before, we had to efficiently manage 32 control accounts. We had to maintain an earned-value accounting system. We had a program management office that stayed busy fulfilling goals.

One of the most significant tasks was validating and rebalancing our basis of estimate. We had to go through it in detail and determine where we were over or under on our proposed budget that was reflected in our Basis of Estimate. We had to consolidate the revised estimates into the baseline of our Earned Value Management System.

We had to understand where we were in the project life cycle. We had to determine whether the client had really done all their due diligence before they went out on contract. Was there some more work that really needed to be done that had been skipped over. We had to make that determination early on if we were going to be effective. We had to choose a management style that allowed for these adjustments. We will get into that a little bit later.

We needed to determine whether we were going to be goal oriented or purpose-driven. Were we going to deliver something that met our goals, hit our budget, and gave us a good score on the earned value management system? Or were we going to listen to the client and make sure we delivered something that was useful in the end.

We chose the latter,

To deliver a solution with value!

Is This What the Client Wants?

A project is truly successful only if it delivers the benefits an organization envisions.

Mark A. Langley
PMI President and CEO

We had to start by reviewing and validating requirements. Every time we did that, or every time that you ever do that you will find deviations from the original requirements.

Collecting those deviations was challenging. In "Honesty Works", Steve Gafney makes this observation about collecting "ambiguous" requirements:

"We often believe that the cost of bringing up an issue is greater than not bringing it up. But is that really the case?

*Studies have revealed the
following statistics: 80 percent of
work problems are due to a lack of
open communications."*

As you collect the evolving requirements, it is extremely important that you keep the contract shop informed. Remember Dr. Ryan's admonishment:

*A stakeholder is someone who has
a <u>RIGHT</u> to influence the system.*

Not everyone with a problem has the right to ask you to solve it. It is important that we elevate issues. Be careful before executing.

If you drop new requirements on contracts, I have found that they will reject them at hand. They generally think that you're just doing a change request for the sake of generating more money. The trick is to ask for change requests that result in additional required benefits. The changes must create value. The value comes from the new b enefits that their stakeholders deem essential. We attacked the program on all four fronts.

Mark Langley, president of PMI said, ***"A project is truly successful only if it delivers the benefit, an organization envisioned."***

Think about that.

- Do you really get a benefits plan when you start a project, or do you get a set of requirements?

- How do you make sure the benefits are successful if you don't know what they were, and you don't understand the why behind those benefits?

- Why is somebody asking us to build the system that they are asking us to build?

Have we lost something in a translation in terms of how the system's going to be used. That was the case with us.

We had to think differently. We decided to follow the four values of the Agile Manifesto.

Think about that. Do you really get a benefits plan when you start a project, or do you get a set of requirements?

How can you ensure that the benefits are achieved if you don't know what they were, and you don't understand the "why" behind those benefits?

How did we get control?

We addressed the project on three fronts:
- Rebalanced the Basis of Estimates

- Documented the difference between Proposal Requirements and Expected Benefits
- Worked with the client's contracts shop to rebalance their budget.

If you go back and look at the Agile Manifesto, you will find direction.

> *Our highest priority is to satisfy the customer through early and continuous delivery of valuable software.*

We need a welcome change. If the customer asks for change, that's good news. That means that we don't have it right. And they want to make sure we get it right.

Welcome Change!

In the earlier, goal-oriented, approach to performance on a program, the idea was to keep the customer away. Don't let them make changes. It was going to throw you off schedule, off budget.

Keep the customer away!

From an agile perspective. You want to welcome those changes. We want the customer to tell us what they expect when this product is delivered.

The concept of "WHY" has been around forever, but let's look at it from the standpoint of PMBOK.

To a degree, the PMBOK has followed suit.

When PMBOK first started in the 2000 edition, it basically said you should <u>keep your client apprised</u> and your management apprised of your progress - progress against the timeline and progress against budget.

You needed to answer questions like:

- Are you on time or are you on budget?
- Will you be ready to release when you're supposed to?
- Were we building something the client wanted?

I guess we can all find cases where that wasn't true.

Subsequent releases of the Project Management Body of Knowledge (PMBOK) slowly evolved to address the fundamental issue.

The Why (Purpose) of a Project Has Grown in Importance

Version	Guidance
PMBOK 2000	Determine Information and communications needs of stakeholders focused on project

reporting and information sharing.

PMBOK 3.0   Manage Stakeholders

PMBOK 4.0   Manage Stakeholder Expectations

PMBOK 5.0   Project Stakeholder Process split from overall project communications - Effectively engage stakeholders throughout the project life cycle, based on the analysis of their needs, interests, and potential impact on progress success

PMBOK 6. O   Benefit Management – A project benefit is defined as an outcome of actions, behaviours, products, services, or results that provide value to the sponsoring organization as well as the project's identified beneficiaries. Development and maintenance of the project's benefits plan is an iterative activity. (PMBOK V 6.0)

PMI added manage stakeholders in version three.

In version 4, we are asked to manage stakeholder expectations. Does that begin to talk about the why behind the project? In version 5 we had to develop a process for stakeholder communication. We are required to keep the stakeholders involved over the entire life cycle.

When you look at version six, development and maintenance of the project's benefits plans became an iterative activity.

Wow! You no longer let the customer build the project benefits plan and write up the charter with requirements while you build a program plan and deliver.

You are put into the position of validating benefits and the underlying requirements. You must understand how these changes are going to affect your original basis of estimate and determine what you must do to rebalance the budget.

Validate your Basis of Estimate (and Rebalance Where Possible)

One of the first things we had to do was to determine how far off we were on the basis-of-estimate that we had bid. We had to be careful not to include the variations that we had already discovered in collaborating with the client. We will manage those in later reviews.

I held these engineering review board meetings every Thursday. As it turns out, this became an every Tuesday and Thursday event.

The various department heads would come in and put their heads down, hopefully not go to sleep, and work out how we were going to balance the budget.

 As a tool, we literally put a pickle jar in the middle of the table and wrote notes for requests and available resources and put them in the jar. If somebody needed additional resources, they would submit a request.

The team would ask, "how many resources do you really need and why? They would either reject the request or if determined viable, put it in the pickle jar.

We would then go around the table and ask, does anybody have an overestimate for something that they had been previously budgeted?

Do we have extra resources that could be re-allocated?

Could we cut corners in a few areas?

We put these new-found resources in the pickle jar as well. This was a very tedious process. Nobody wanted to give up their budget and everybody wanted more.

But,

it was highly effective because it made us realize that in delivering what was bid at $212 million required that we had to be very, very frugal.

We had to adhere strictly to the requirements as they were written as part of the procurement.

I knew that every time that we visited a stakeholder, they would say something like, "that's not what I wanted."

That became frustrating.

You wanted to tell them, No!, that's not what we bid. That was not included in the requirements documents. Your organization didn't buy that capability.

The stakeholder heard: "We'll give you something. You may not be able to use it, but that's what you bought".

Where Were We in the Project Lifecycle?

For that matter, where was the customer in the project lifecycle? Had they completely thought out the details well before competing and issuing a contract?

What Steps were Skipped?

Had they started with a well thought out project charter or had they just responded to a somewhat disorganized set of complaints (requirements)?

If you believe the PMBOK, the stakeholders generate the needs assessment, the benefits plan, business case, and then the project charter.

Did we start with a well thought out Project Charter?

With the charter in place, the contractor (us) becomes involved by building the project plan, following the plan, and delivering the solution.

As we discussed in an earlier chapter, version six of the PMBOK states that the benefits management plan is iterative, not fixed. If you are running a goal-oriented project, you are best served by minimizing changes to the benefits plan and following the traditional waterfall approach.

On the other hand, if you are purpose-driven, you put the benefits plan in the forefront, realizing that it is an iteratively developed document. As you go through each iteration, purpose is better defined, and additional benefits are identified.

This opens an exceptionally large can-of-worms. The client had obviously skipped a few steps and obviously did not want to be reminded of the gaps in requirements.

I like the way Mitre lays out the project lifecycle. Mitre has a more overarching point of view. They say, okay, let's talk about the overall lifecycle of a project.

- Where does it start?
- Where does it end?
- They realize that you must first develop some broad project scope.
- Next, you hone that down and get broadly covered requirements.
- You collaborate with the stakeholders back and forth, to develop a detailed, validated set of requirements.
- You do a little more work, prioritizing those requirements.
- With a little more work, we have documented requirements.

At that point in time, if all has been done well, you select a contractor to do the work and turn them loose to complete the contract.

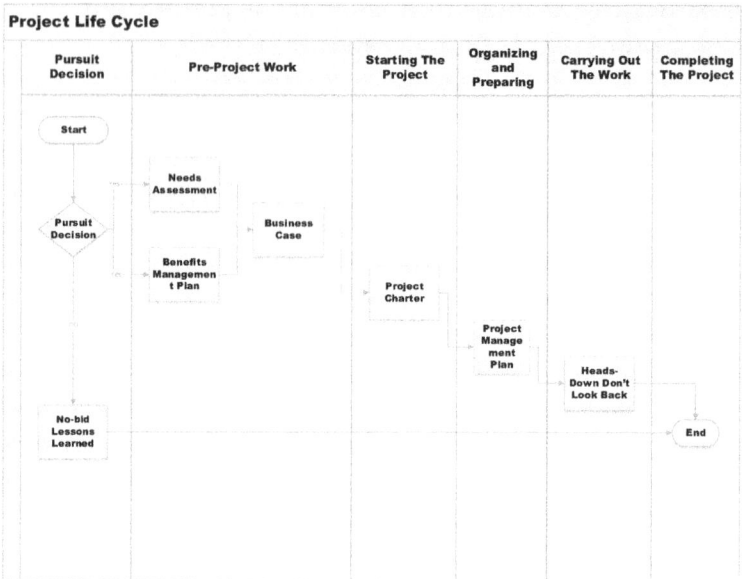

Project Life Cycle

Pursuit Decision	Pre-Project Work	Starting The Project	Organizing and Preparing	Carrying Out The Work	Completing The Project
Start					
	Needs Assessment				
Pursuit Decision	Business Case				
	Benefits Management Plan				
		Project Charter			
			Project Management Plan		
				Heads-Down Don't Look Back	
No-bid Lessons Learned					End

I don't know about you. I've been doing this for a lot of years, and I don't think I've ever been given a contract to do something where all these steps were taken.

Contract Awarded Assuming That Requirements Gathering was Complete.

You have a contract in hand with a set of requirements. You have bid a fair price to execute the work assuming that the requirement gathering process was complete.

The contracts office assumed that the necessary steps had been taken. The requirements were all covered, validated, prioritized, and documented. You soon discover that the requirement gathering process was not complete.

If you are purpose-driven, you call a pause and start the contract by reaching out to the stakeholders to validate their requirements.

As you first meet, the stakeholder might state,

We gave you the requirements.

Why are you asking me to explain what I am trying to accomplish?

As the discussion evolves it becomes evident no one completely understands the requirements. At least not sufficiently to implement the solution.

The client will go on to say, "you should know what I am trying to accomplish". You should know the benefits that I will enjoy when you deliver."

The client was expecting that we would deliver the solution as it was outlined in the contract.

If all the validation steps had been concluded as required, we would just put our heads down and build. The following chart is based on Mitre's recommended approach to establishing project requirements.

The client believed that we should know what they are trying to accomplish. They expected us to know the benefits that they will enjoy when the system is delivered.

We had to take a step back and work with them to define and understand the expected benefits. We needed to know "why" the system was being built.

That is what we did. We had to pull those requirements out of the contract and out of the stakeholders and then institutionalize them and make them part of the design doc.

Back Up and Collect Requirements

We were hoping that the requirements had been detailed and all we had to do was do some prioritization.   I'll take those, and then we'll prioritize and work through them. Build a series of sprints, build a backlog and start working that backlog off.

The underlying problem was that the contract was issued with very broadly described requirements. The requirements did not have enough detail to completely identify the clients' needs or the operational constraints.

We were facing real problems. At this point in time, we were about 90 days into the contract, and we needed resolution.

I held a meeting with the key stakeholders, the clients, and their contract shop.

I told them, without pulling punches, that we could build exactly what we bid for $212 million. I felt confident that we could do the job as we had bid.

We had gone through and looked at our basis of estimate. We had wrestled through that level of detail. We had worked to compromise. We found areas where we could give a little to offset the takes.

If they wanted a solution that prioritized value, we would need to back up and do some additional work.

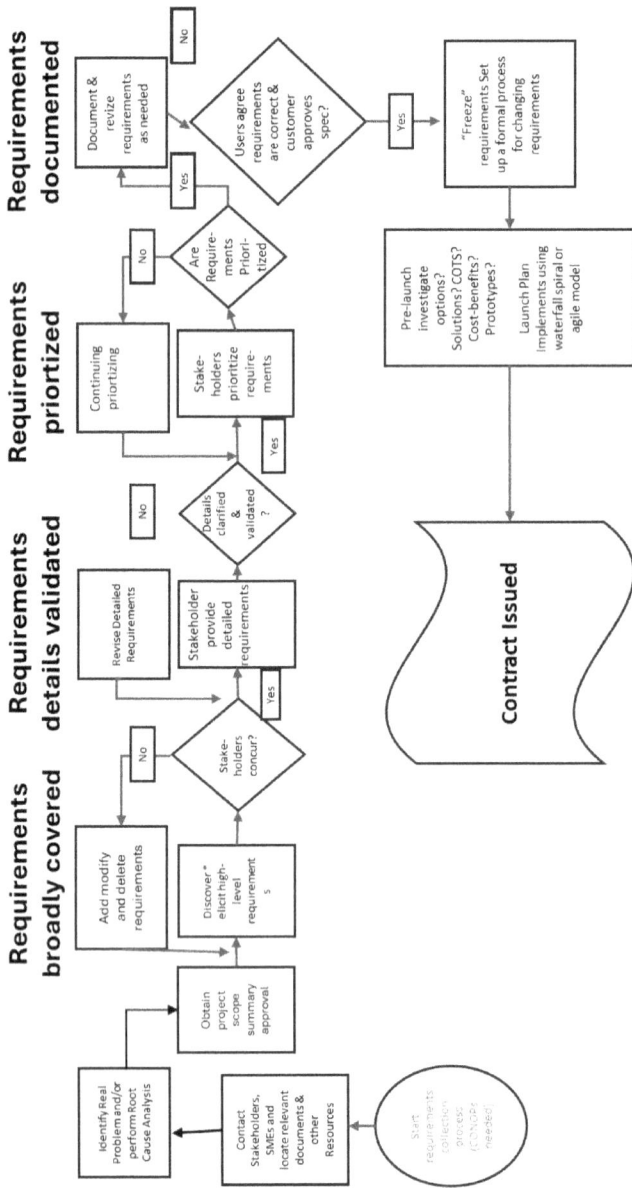

Requirements documented
Requirements prioritized
Requirements details validated
Requirements broadly covered

Document & revize requirements as needed

Users agree requirements are correct & customer approves spec?

No

Yes

"Freeze" requirements Set up a formal process for changing requirements

Are Requirements Prioritized

No

Yes

Continuing priortizing

Stakeholders prioritize requirements

Yes

Pre-launch investigate options? Solutions? COTS? Cost-benefits? Prototypes?

Launch Plan Implements using waterfall spiral or agile model

Details clarified & validated?

No

Revise Detailed Requirements

Stakeholder provide detailed requirements

Yes

Contract Issued

Stakeholders concur?

No

Add modify and delete requirements

Discover* elicit high-level requirements

Yes

Obtain project scope summary approval

Identify Real Problem and/or perform Root Cause Analysis

Contact Stakeholders, SMEs and locate relevant documents & other Resources

Start requirements (collection) process (COMBPr needed)

Client Believed That They had Completed Each Step

Prioritized Value ("Why")

We needed to add the following corrective steps to ensure that they would realize the necessary value.

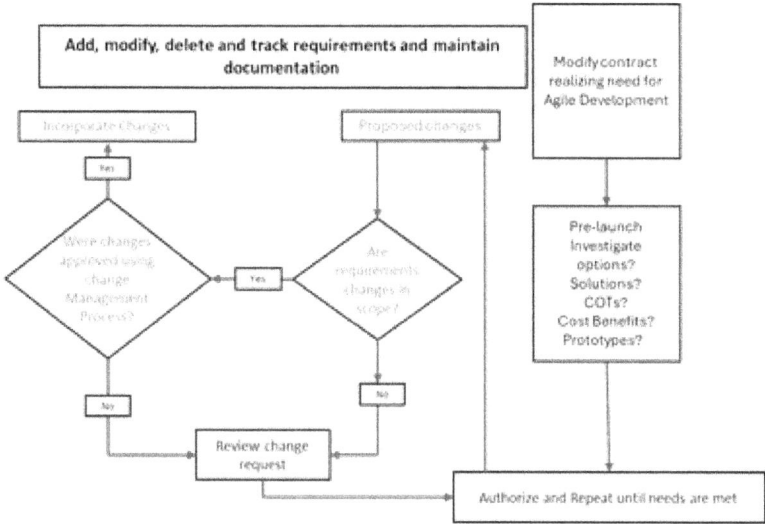

It became obvious when talking with the actual stakeholders, that there was a lot of stuff left out of the procurement- the details, (the devil in the details), adding more work.

We were going to have to reevaluate if they wanted to incorporate this added information.

We could build it for what we bid for $212M, and come back later to satisfy these clarified requirements, or we could try and build the right thing and build it the right way.

The client chose the latter and modified the contracts to include collecting and validating the requirements and building a system that would provide the benefits that were necessary to meet the user's needs.

Revised Contract Issued for Over $300M.

A revised contract was issued for more than $300M. We took a deep breath and found out what was really needed.

The sponsor for a project is often unaware of all the stakeholder's needs. If you are a purpose-driven project manager, you should do what it takes to do the right thing.

> *"The project sponsor is generally accountable for the development and maintenance of the project business case document." ..." A needs assessment and benefits plan precede the business case."*
>
> PMBOK v6

Does the sponsor always communicate the Purpose? Do they provide:

"The WHY of the Project"

We found that the project sponsor doesn't always communicate "why" behind their project. They kind of gloss over that.

It takes a lot of challenging work to understand the WHY. We found that we had to look at the benefits plan as an evolving document. The more we did, the more we found.

In his book, *"Dilbert 2.0: 20 Years of Dilbert"*, Scott Adams had this pegged. After a while, you conclude that:

They want you to tell them their requirements.

Dilbert's last frame sums the client's position: "Can you design it to tell me my requirements?"

© Scott Adams, Inc./Dist. by UFS, Inc.

I can think back over several decades and realize that I didn't usually have the luxury of working with a benefits plan. Even when I asked for it, it was very limited. What was worse was that quite often the customer or the client had not even developed a benefit plan.

They just assumed that there was a gap in operations, and they were not certain how big or where that gap was hiding. They just assumed that the resulting solution was going to fill the gap. They assumed that their needs would be met and value delivered.

They haven't answered the "WHY" of the project.

"All project management methodologies demand that projects have a well-defined business case. But when evaluating and prioritizing projects, looking at the business case alone is not enough. We also need to understand how the project connects to a higher purpose. Evaluating on purpose can help leaders decide whether the project aligns with the organization's strategic goals. It is also a key driver for engaging team members and the organization as a whole and motivating them to support the project.

Companies must learn how to articulate a project's purpose. A straightforward method of uncovering it is to simply ask, "Why are we doing the project?" Then, when you've arrived at your answer, ask why again. With each successive layer of why, you will come closer to the project's purpose. If this exercise does not help you discover something that will motivate people to work on the project, you probably should not start it.

Antonio Nieto-Rodriguez, HARVARD BUSINESS REVIEW PROJECT MANAGEMENT HANDBOOK

The important thing in life is to have great aim and to possess the aptitude and the perseverance to attain it.

Johann Wolfgang Von Goethe

Chapter 10 - Impact of Selecting Purpose-Driven Approach

If we go back to earlier versions of the PMBOK, we would be tempted to follow a Goal Driven Paradigm. In later versions we learned that it is important to understand the "why" of the system.

Goal Driven Paradigm

PMBOK 2000 edition

10.1 Communications planning

- Determining the information and communications needs of stake holders, when they will need it ,and how will it be given to them.

10.2 Information distribution

- Making needed information available to project stake holders in a timely manner.

10.3 Performance reporting

- Collecting and disseminating performance information. This includes status reporting, progress management, and forecasting.

10.4 Administrative closure

- Generating, gathering, and disseminating

information to formalize a phase or project completion.

Purpose-Driven Paradigm

PMBOX V6

13.1 Identify stakeholders.

- The process of identifying project stakeholders regularly and analysing and documenting relevant information regarding their interests, involvement, inter dependencies, influence, and potential impact on project success.

13.2 Plan stakeholder engagement

- The process for developing approaches to involve project stakeholders based on their needs, expectations, interests, and potential impact on the project.

13.3 Manage stakeholder engagement.

- The process of communicating and working with stakeholders so that their needs and expectations are addressed. Identify issues and foster appropriate stakeholder engagement involvement.

13.4 Monitor stakeholder engagement

- The process of monitoring project stakeholder relationships and tailoring strategies for engaging stakeholders through the modification of engagement strategies and plans.

If we go back to the paradigm of being goal-driven that we find in the earliest PMBOKs, the whole point was to just inform the customer how we were progressing on the project. If they gave us a charter to conduct, we put together a program plan, followed the plan, and delivered.

We were going to keep them informed on a regular basis. We would tell them how we were proceeding against that plan?

By the time we get to version six, PMBOK started talking about talking **with stakeholders**, **not to them**.

Talk with Stakeholders, Not at Them!

Now we must have a process for communicating and collaborating with stakeholders to make sure that we meet their needs and expectations.

We must understand **why** we're building this system. That's what it says in PMBOK. Unfortunately, The PMBOK doesn't talk much about mapping requirements against benefits, even less about mapping requirements to the **WHY**.

In 2015, PMI published a Global Standard, "Business Analysis for Practitioners", that provides guidance on needs analysis. Their proposed approach helps to answer the "WHY" question.

I think that's imperative in building effective solutions. You must include a detailed needs assessment.

Needs Assessment

According to "Business Analysis for Practitioners", needs assessment work is undertaken before program or project work begins. It is considered to be a pre-project activity. I have found that it is an ongoing activity.

As you follow the Agile process, you will find that needs evolve. They are never completely defined. You will need to include recurring needs assessment if you are to build a valued solution.

The "Business Analysis" guide is a good companion to the PMBOK in an agile environment. It helps guide you through the needs assessment phase that will likely arise as you build solutions.

Needs assessment can be broken down into four steps:

1. Identify Stakeholders
2. Assess Goals and Objectives
3. Recommend Action
4. Update the Business Case

Identify Stakeholders

Common techniques that can be used in the discovery of stakeholders are brainstorming, decomposition modelling, interviews, surveys, or organizational modelling.

As you identify the stakeholders, you need to understand their relative position. Are they responsible, accountable, consultative, or strictly informative. I built a RACI chart.

	Sponsor	Product Manager	Business Analyst	Product Development Team	Project Manager
Identify problem	A	C	R	C	C
Address current state	A	I	R	C	C
Recommend action	I	A	R	C	C
Update business case	I	A	R	C	I

Remember, not all stakeholders have the authority to authorize a change. Elevate needs but wait for authorization before jumping off the cliff with a solution.

As we iterate, collecting more information, additional needs will be identified.

Remember the forms I used in documenting these new needs? The needs were generated by stakeholders and adjudicated.

Current State

We need to determine how the requested "need" interacts with the current state.

I remember a case where a stakeholder wanted to use multiple data sources in developing a predictive analytical tool. That sounded great until I found that one of the data sources was highly restricted and the stakeholder did not have authorization.

The current state prohibited implementation.

Recommend Action

The recommendation can be as simple as requesting a mod to the project plan. Or it could be as complex as modifying business rules.

Update Business Case

The resulting needs must be formalized, documented, and authorized before commencement of the solution.

Remember to keep contracts informed.

We need to understand the constraints on the management approach that we will follow in building these evolutionary solutions.

Choose A Management Style

Should we go with a goal-oriented approach, or hopefully by now, we have decided to follow a purpose-driven approach.

Your management style is dependent on your mindset. Dr. Carol Dweck talks about mindsets in her book by the same name: *"Mindset: The new Psychology of Success"*.

You have two choices; you can live with a Fixed Mindset, or you can expand with a Growth Mindset. If you follow the pure goal-oriented path, you are more fixed. However, the purpose-driven approach is more growth oriented.

Your mindset governs your style of project management.

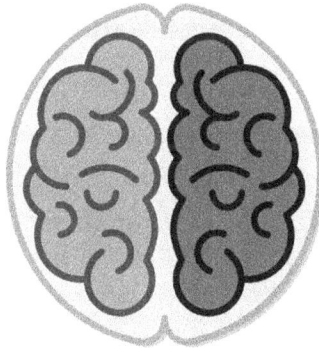

Growth Mindset	Fixed Mindset
Embraces challenges	Avoids challenges
Accepts criticism and	Rejects criticism and is

negative feedback as constructive	hurt by negative feedback
Equates reward with effort	Expects reward without effort
Persists in face of setbacks	Let's setbacks derail them
Never give up	Gives up easily
Learns from failure	Unable to accept failure or mistakes
Talent is developed	I'm no good at this. Talent is static.
What more can I do?	Why should I bother?

Honestly ask yourself, "what is my mindset?", then settle on you project management style.

Goal Oriented Project Management

"The purpose of project management is to foresee or predict as many dangers and problems as possible; and to plan, organize and control activities so that the project is completed as successfully as possible in spite of all the risks."

Burek, Paul (2008)

Purpose-Driven Project Management

The purpose of project management is to make the customer successful by identifying the underlying needs of the client and satisfying those root causes

(Success is not about just meeting requirements, it's about fulfilling the "why" behind requirements.)

We must remember to keep purpose ahead of the goal. We decided that we wanted to make sure that what we were delivering to the customer was going to be useful. At the same time, we wanted to make sure that we stayed on budget.

If we were going to do that successfully, we needed to carefully validate and refine requirements. As you are going through requirements, new ones are going to come out that you had not expected.

As you go through the steps illustrated in the Mitre approach discussed in the section on validation of your basis of estimate, you must back up and do that detailed validation, detailed prioritization, and detailed documentation. This will provide new opportunities to provide value and fulfill the clients' needs.

When you do this, you're going to expand the scope of this contract. That is the mess that is best exposed in the beginning of the project. You are likely to find too much implied scope and not enough money. Be careful not to break the bank.

Scott Adams caught this one back in 2001. (from *"Dilbert 2.0: 20 Years of Dilbert"*)

Dilbert saw the increased number of requirements as an increase in complexity. I see these evolving requirements as a better understanding of purpose.

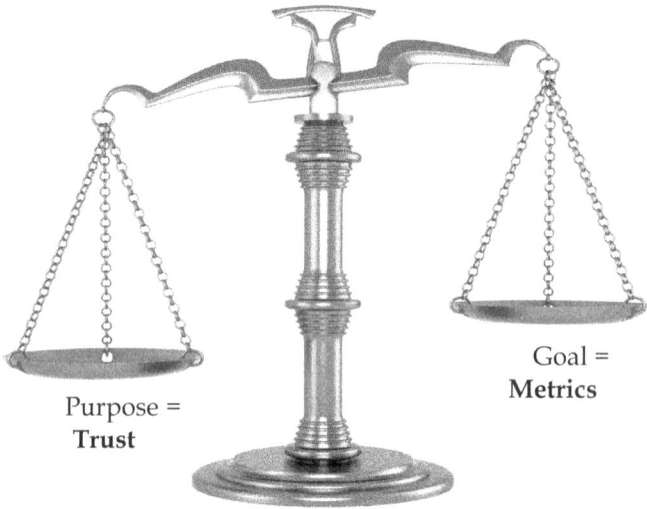

Purpose =
Trust

Goal =
Metrics

Always keep Purpose Ahead of

Chapter 11 - Scope Will Creep

How do you quantify implied scope?

I had an excellent system engineer working for me that shared a ratio that Lockheed Martin used.

He said, if you start with a hundred top-level requirements, they would probably yield about 500 design requirements.   I kind of pushed that aside a while, but as we were building requirements, I found that he was right. We started with one number. At the end, we had five times as many requirements in the design documents.

The key was to document this evolution of client requirements.

When we sat down with the stakeholder and they said, "you did not get it right." This is what we really mean. This is what we really need. We made sure that the person that was working with the stakeholder documented the interaction.

- What was the original requirement?
- How did it change?
- Was it a new requirement?
- Was it a modification of an old requirement.
- How did it impact the project plan?
- How did it affect the Benefits Plan?

We had a form printed out that all my people carried around with them.

Update Requirements and Align with Benefits

When anybody asked for a change, we wrote down exactly what needed to be changed or added.

- How it related to the existing contract.

- How would the change provide benefits to the resulting system.

- Why was it of value?

We then asked the stakeholder to cosign the form. We asked them to put their name down as the person that identified this issue.

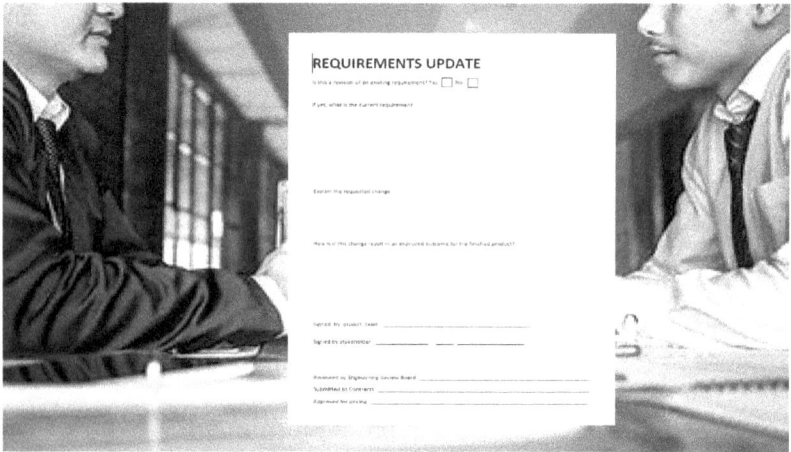

Keep the Contracts' Shop Informed

About once a week, I went to see the contract shop and I gave them this stack of completed forms.

I said, I am not saying I'm going to do any of this.

I'm just trying to say, this is what your stakeholders say that they need.

I don't know where it got lost in translation.

I'm sorry, it did, but this is what they want.

You tell me, which of these do you approve and get these back to me and I will develop an estimate that we could include in our validated requirements, project plan, and in the benefits plan.

After approval, we will modify the baseline of our value management system.

We will get it built for you.

We will build what your stakeholders want, not just what the contract specifies. Not everything changes. The traditional life cycle remains somewhat linear in versions six and beyond.

Traditional Life Cycle Remains Somewhat Linear

Project Life Cycle

Pursuit Decision	Pre-Project Work	Starting The Project	Organizing and Preparing	Carrying Out The Work	Completing The Project

Start

Needs Assessment

Pursuit Decision

Business Case

Benefits Management Plan

Project Charter

Project Management Plan

Heads-Down Don't Look Back

No-bid Lessons Learned

End

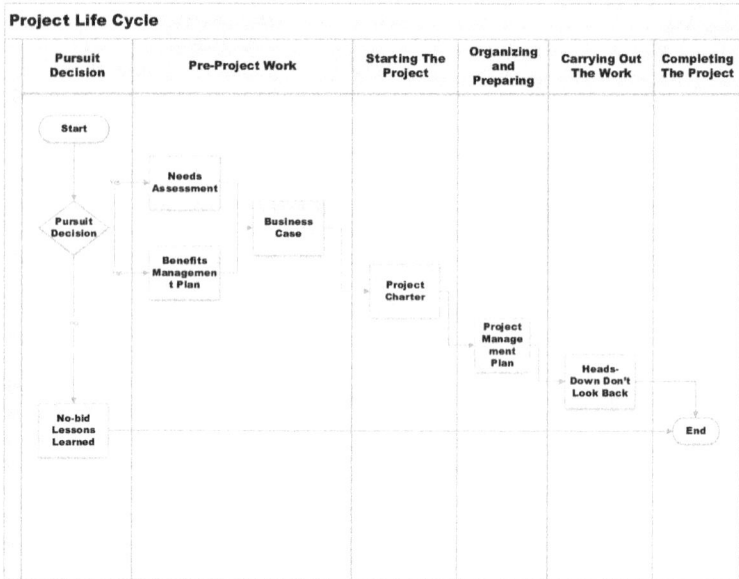

It is Not so Predictive.

As discussed previously, most systems are not that predictive. As you collect requirements and even as you build the system, you will uncover additional pain points that must be resolved.

These new pain points usually require that you modify the benefit plan, perform additional analysis, and do some redesign.

Benefits are Evolutionary.

I like the way Isabelle Aguilera discussed this in a paper entitled: "Delivering Value" published by PMI.

She asked meaningful questions.

- Have the benefits been developed and agreed upon?
- Have key stakeholders been consulted?
- Are the reporting structures and benefits monitoring process in place?
- Have the benefits been reviewed and updated?
- Are benefits aligned to the organization's strategic objectives?

Boy, I wish I'd have known Isabel when I did this project. I had to figure some of these things out for myself. Check out that paper "Delivering Value" by Isabelle Aguilera published by PMI. I think that you will learn a great deal.
She went on to say that project managers should be empowered to identify, approach, and fix as many areas as possible.

> *"Project management should be empowered to identify, approach, and fix as many areas as possible – including strategic ones – which includes taking a role in monitoring and measuring the benefits those projects deliver to the business."*

> Isabel Aguiler
> Delivering Value PMI

Keep these points in mind as you work on refining requirements:

- One hundred top level requirements will probably yield five hundred design requirements.
- You must document client requests, write them down.
- Have the stakeholder co-sign with you. (If there is not an artifact, it did not happen)
- Keep contracts in the loop.
- Maintain the project plan to reflect the changes.
- Maintain the benefits plan.

You're not dealing with a small narrow view of a problem. You are looking at the wide scope of the problem and how can you include all of this into the strategic view of the system you're building? How do you relate to the environment that you are working in? You need to deliver business results.

The LinkedIn AI blog on IT/Services/Software Management reminds us that we must deal with the dynamics of a project.

Embrace Flexibility

If you check out the Benefits Realization Management (BRM section 2.4.4), A Practice Guide published by PMI you will find more emphasis on understanding "WHY?".

2.4.4 EMBRACE FLEXIBILITY

Organizations that tend to be good at BRM embrace flexibility, because in dynamic and changing environments, few programs or projects are delivered as originally planned. This implies that either BRM may not deliver the planned benefits, or that expectations change during the delivery timeline. As the delivery environment changes, the benefits being sought may also need to be changed/adapted.

Benefits Realization Management
A Practice Guide
Project Management Institute

Organizations that tend to be good at BRM embrace flexibility. They don't put up barriers when listening to the client.

Because, in dynamic change environments, few programs or projects are delivered as originally planned. This implies that either BRM may not deliver the planned benefits, or the expectation has changed during the delivery against the timeline.

Th expected benefits (value) may change. Remember the Space Command story? You must be able to change and adapt.

PMI goes further in the AGILE Practice Guide. It states that the benefits plan should also be maintained in an iterative life cycle. We must do that if we are able to deliver value in terms that the client understands.

Staying organized also means being adaptable and responsive to changes and challenges. Review and adjust your project plan, scope, goals, methodology, tools, and communication regularly, based on data, feedback, and lessons learned. Monitor your project's performance, quality, risks, and issues, and take corrective actions when needed. Involve your team and stakeholders in the review and adjustment process and document the changes and outcomes.

LinkedIn Blog

Iterative Development Approach

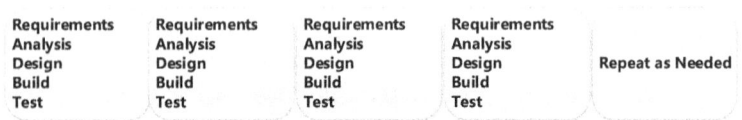

| Requirements Analysis Design Build Test | Requirements Analysis Design Build Test | Requirements Analysis Design Build Test | Requirements Analysis Design Build Test | Repeat as Needed |

Iteration-Based Agile
"Agile Practice Guide"

A Purpose-Driven Iterative Based Life Cycle kept the Benefits Plan Up to Date and allowed us to be Purpose-Driven.

The Agile Practice Guide published by the Project Management Institute in cooperation with the Agile Alliance, reminds us that we must add needs assessment, benefit analysis requirements, and design to each iteration until you fulfill the client's need. Once you have completed those steps, the requirements and resulting benefits can be updated into the backlog. As the sprints are worked off in an agile process, the benefits will be realized.

This $212 million dollar program turned out to be a $390-million-dollar engagement.

As it turns out, one size doesn't fit all. There is a continuum of life cycles, some predictive, some more suited to an agile approach.

The parts of the system that were predictive in nature could be delivered incrementally. We found that some parts were more iterative and had to be delivered using an agile approach.

We used the agile side of the house to outline capabilities. The infrastructure system and function development used an incremental approach.

We used a concept visualization technique to expose the Operational Activities that were required.

Those concept visualizations were expanded into use case diagrams to define how the Operational Activities were conducted, how they were supposed to work, and what functionality was required to support that required operational capability.

We used the DoD architectural framework to map this all out. We literally had a big room to display both the system and operational views.

On one wall we displayed the system views that defined the systems and their corresponding functions. On the other side were all the capabilities and their operational activities.

When we exposed a required operational activity, we would look over at the system views on the other wall and identify the system and function that was going to support that required operational activity.

If we didn't have that function, we would modify one of the systems or add a new system to the architecture that would fulfill it or provide the data necessary to fulfill it.

Several people from the office of the secretary of defense came down to look at our process and really loved the approach. We were using DODAF to build a system that was relevant. It was one of the most exciting parts of this contract.

I don't expect everybody to follow DoDAF views. It's a very, very detailed approach, but if you read the DoDAF version 2.02, I think you'll find a lot of good ideas on how it marries program management with solution architecture. It helped me to understand how you can balance these hybrid type of projects, part predictive and part iterative.

The Continuum of Life Cycles

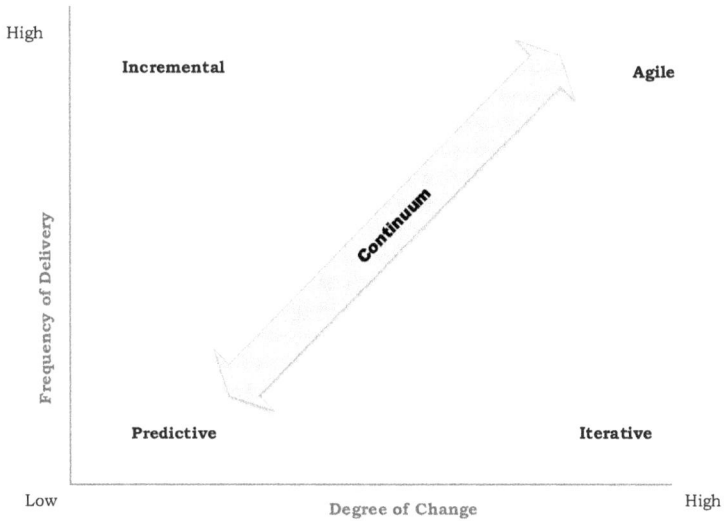

We used both capability (CV) and system (SV) views to document the eventual solution.

Capability Views

CV-1: Vision	Addresses the enterprise concerns associated with the overall vision for transformational endeavors and thus defines the strategic context for a group of capabilities.
CV-2: Capability Taxonomy	Captures capability taxonomies. The model presents a hierarchy of capabilities. These capabilities may be presented in context of a timeline - i.e., it can show the required capabilities for current and future capabilities.

CV-3: Capability Phasing	The planned achievement of capability at different points in time or during specific periods of time. The CV-3 shows the capability phasing in terms of the activities, conditions, desired effects, rules complied with, resource consumption and production, and measures, without regard to the performer and location solutions
CV-4: Capability Dependencies	The dependencies between planned capabilities and the definition of logical groupings of capabilities.

System Views

SV-1 Systems Interface Description	The identification of systems, system items, and their interconnections.
SV-2 Systems Resource Flow Description	A description of Resource Flows exchanged between systems.
SV-3 Systems-Systems Matrix	The relationships among systems in a given Architectural Description. It can be designed to show relationships of interest, (e.g., system-type interfaces, planned vs. existing interfaces).
SV-4 Systems Functionality Description	The functions (activities) performed by systems and the system data flows among system functions (activities).

What was left over was a cross-reference diagram. That diagram is called a system view, five-B (SV-5b) in DoDAF parlance.

That view lines up the various operational activities that were needed and the requisite functions that are necessary to carry out those activities.

		Capability 1		Capability 2		Capability 3	
		Activity A	Activity B	Activity C	Activity D	Activity E	Activity F
System 1	System Function A	R		R			R
	System Function B				G		Y
	System Function C		R			R	
System 2	System Function D			G			
	System Function E				Y		Y
	System Function F						Y
System 3	System Function G		G				
	System Function H		G		R		
	System Function I		Y				

DoDAF SV-5b

The green, yellow, and red bullets indicated how much we had to stub out.

The Reds are not there yet, more work is needed. The greens are what's there today. They are the system functions that are there today that we could show and demonstrate to support the specific operational capability. The yellows were the ones that were in process.

We could work the yellow intersections together.

I have included a description of other service and capability models in Appendix D. I have found the operational views (OVs) helpful in better understanding the client's needs and answer the "WHY?" questions.

This is a simplistic view, but when you have as many activities as we had, remember, we ended up with hundreds of requirements, it became more complex. It gets tedious, but trust me, it worked quite well. When you need to explain progress to a stakeholder, you can use this view to help them visualize a particular capability or operational activity that's going to be available in the finished system. Where appropriate, you can stub that out.

Concept visualizations of those stubs can help in the communication of a common understanding of the purpose of the system.

Even after delivery, sometimes long after delivery, requirements and benefits continue to morph.

Requirements and Benefits Continue to Morph.

The Space Command in Colorado Springs is required to satisfy many requirements. The command must support the varied and ever-changing world we live in.

As is the case in many systems, even after delivery, requirements and benefits continue to morph. Here is a quick story about requirements that had not been validated and the question of "WHY" had not been asked.

A couple of decades back, I chaired a requirements analysis workshop for the Space Command out in Colorado Springs.

We had a team on our panel that went out and identified fifty open requirements that were being maintained by the CIO of space command. We drove those requirements to the ground and found out exactly what the "WHY" was behind each of those requirements.

- Who expressed that requirement.
- Who was the stakeholder?
- Does that person still exist?

- Does that organization still need that requirement?
- Is that still a valid requirement?
- Is the WHY still there that requirement was meant to fill?

I picked a simple one to start.

Objects in orbit are monitored. Orbital mechanic specialists have developed algorithms (I worked with eight of them) that predict the location of these objects to avoid potential collisions.

The Maui observatory checks on the objects orbital locations as they pass over Hawaii. If the object is where the algorithm predicts, all is well. If not, adjustments must be made.

The data from the Maui sensors is transmitted to Cheyenne Mountain in Colorado Springs.

That seemed straight forward until I reviewed an open requirement. The research team (Lincoln Labs in Massachusetts) asked that Space Command capture all the transmittals of all the orbital data. This was just in case the algorithms were defective and numbers had to be backtracked.

I visited the mountain and asked about that open requirement. An airman walked me over to a printer that was typing out the data coming from Maui.

The box of paper was about to be empty, the dutiful airman opened a new box and loaded the printer.

I asked, what are you going to do with the printouts? He offered to show me.

We left the building (inside a cave in the mountain) and walked to a nearby cavern that was full of steel shelving.

The airman dutifully added the new box to an ever-growing collection of boxes.

No one ever came for the boxes, and no one ever needed the data. As it turns out, the request for collection was only intended to cover the startup period, not for several years of operation.

One down, forty-nine to go!

Guess what? Out of fifty requirements, only two continued to be valid requirements that were needed at that point in time, forty-eight were requirements that were no longer needed. Yet, these fifty requirements had been maintained on the backlog of the CIO. They were shown as open requirements or deficiencies in the current system.

Before briefing an assemblage of over one thousand contractors and DoD personnel, I thought it best to brief the CIO of Space Command.

I am glad that I did. At first, he was upset with the audacity of my report. After walking him through the process, he calmed.

I reminded him that the "unfulfilled" and "unnecessary" requirements went on the books long before his arrival and that our team felt that he would have found the same information in short order.

He concurred and thanked us for our work.

I think you may find this is not uncommon. Always remember to validate requirements before you build and recheck them as you implement the solution. Also, ensure that uncovering those requirements does not embarrass the client.

This is much easier to remember if you follow a Purpose-Driven approach. Ask "why" we are doing this. Why do we have this requirement?

Requirements come in many flavours.

Chapter 12 - Five types of Requirements

 As you unfold a project and in a previous section, I mentioned the work of Noriaki Kano. Professor Kano developed an interesting concept that should be considered as we collect and define requirements. In 1984, Professor Kano proposed that there were five types of requirements.

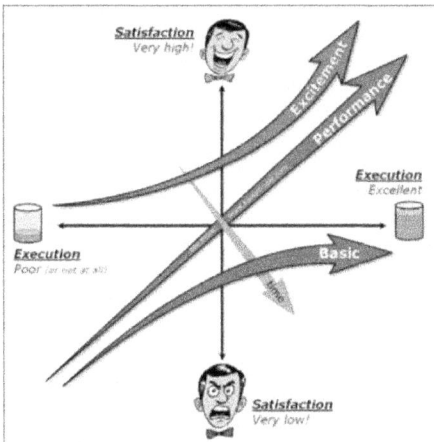

- Performance
- Basic
- Excitement
- Indifferent
- Reverse.

The Kano Model

Prof. Noriaki Kano created the Kano Model in 1984

User Experience Magazine 1999

Performance

These are the requirements the customers can articulate and are at the top of their minds when evaluating options.

They are such things as

- throughput capability
- delay time on transmission of information,
- refresh times on screens.
- bandwidth,
- ETC.

These are usually easy for clients to give you a contract start. They are usually found in the system specification, not in the requirement docs. So that's performance.

Basic

These are the requirements that the customers expect and are taken for granted. When done well, customers are just neutral, but when done poorly, customers are very dissatisfied. Kano originally called these "Must-be's" because they are the requirements that must be included and are the price of entry into a market.

These are the things like, "it must be able to address these five databases". I don't care how you do it.

It's not going to tell you that one of those databases is highly restricted and only a few people are even allowed to look at that data.

So, they may have asked you a question to amalgamate data from five different data sources. But suddenly, when you deep dive into the requirement, you find that the general people who use this capability only have access to four of them.

Only a small group of people have access to all five.

That's the kind of problem you have when you have basic requirements. The client thinks that these are simple capabilities to provide. You may also assume simplicity until you start trying to understand the restrictions. These discovered requirements become exciters.

Excitement

These are the requirements that are unexpected and pleasant surprises or delights. These are the innovations you bring into your offering. They delight the customer when there, but do not cause any dissatisfaction when missing because the customer never expected them in the first place. Kano originally called these "Attractive or "Delighters" because that's exactly what they do.

These are the requirements that you find in working with the stakeholders that they had never really written down.

These become extremely important when exposed. So much so that they become delighters. They are what makes customers excited. That's what makes them really build a trust level with you because you're seeing things from their point of view. You're trying to make the solution work from their point of view, not from yours.

Indifferent

These are the requirements that the customer simply doesn't care if they are present or absent, their satisfaction remains neutral under either circumstance.

Indifferent are those requirements that the RFP drafter put in, but no one knows why. Somehow, they got into the requirements document, but nobody can understand why in the world we have them. (Remember the Space Command Story?)

- They really don't provide any value.
- Nobody can map them up to a benefit.

Quite often, these are good candidates for trade-ins. If you need more money, tell the client how much money they were going to pay to solve this indifferent requirement?

Do you want us to not do it? The contracting officer can understand that and say, okay, that puts more money in the pot. In my case, it put more money in the pickle jar so that we could find other ways to solve problems.

Reverse

These are the requirements that cause dissatisfaction when present and satisfaction when absent.

Reverse Requirements are dangerous. These can be gotcha's. If you implement this capability, it's going to cause a problem. It's going to stop something else from working. It's going to break the rules.

If a client has a set of rules about privacy of information, and you suddenly implemented something great that provided a lot of data to the end-user, but what you have created broke privacy rules. You have real problems because of the privacy rules in today's world. Those are rules that you must not break.

Watch out for those as well.

"As we discover the new Delighters, development and maintenance of the project's benefits plan becomes an iterative activity, resulting in new requirements."

Over time, these newly identified requirements that are initially delighters, become basic needs. You've pointed out something that stakeholders didn't even know they needed.

You have created a "Must Have".

It's something that must be there for the system to provide the desired benefit. It is important to capture those newly created basic needs contractually and where necessary, adjust the contractual agreements.

Your brilliance can be very costly. You may find yourself projecting your vision onto the clients' problems. It may be something that you value, but not necessarily something that the client values.

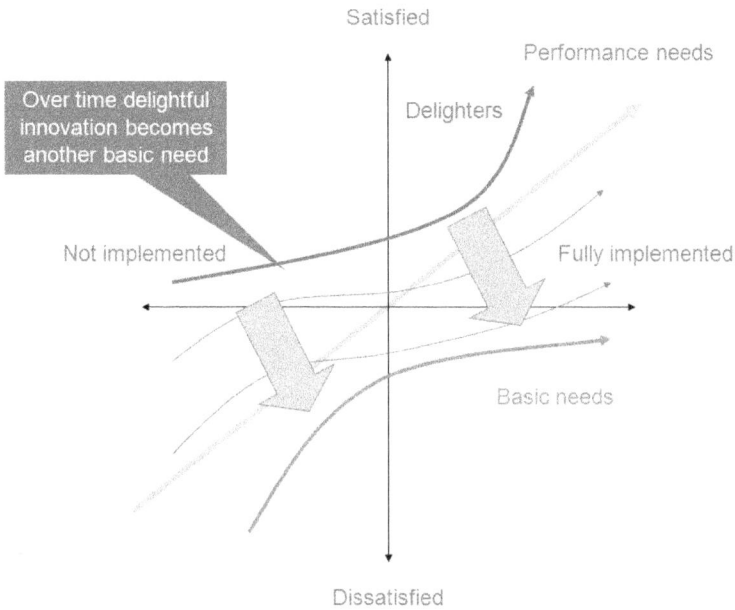

It is very easy to agree to a change and increase scope without much effort. Unfortunately, you may not get paid for that extra effort. It is important that you keep contracts involved.

Contracts personnel are a different breed. They are very goal oriented. Purpose is generally not part of their vocabulary. They think, dollars, deliverables, and schedule.

The best way to get them in your corner is to communicate value add and when necessary, the loss of value without a modification.

Enlist your stakeholders to document the potential value before doing battle with contracts.

In his book, "The Lean Produce Playbook.", Dan Olsen discusses how understanding and exploiting the different types of requirements can result in a more valued solution.

Make sure that the client's must-haves are fully addressed. Don't make the mistake of leaving them out.

Make sure that you can exhibit performance in terms that the client understands.

Highlight the delighters that you have identified. They may not even appear in the solicitation, not in requirements or even specs. Delighters become discriminators.

Dan must have read some of Professor Cano's work.

Example of Completed Product Value Proposition			
Competitors	A	B`	C
Must Haves			
- Must-Have 1	Yes	Yes	Yes
- Must-Have 2	Yes	Yes	Yes
- Must-Have 3	Yes	Yes	Yes
Performance Benefits			

- Benefit 1	High	Low	Medium
- Benefit 2	Medium	Medium	High
- Benefit 3	Low	Medium	High
Delighters			
- Delighter 1	Yes		
- Delighter 2			Yes

Dan is very Purpose-Driven!

In a competitive situation, these concepts can separate you from your competition. The trick is to propose and deliver a better "Value Proposition."

Concept visualization is a straightforward mechanism for eliciting needs and establishing value.

*Ask, and it shall be given to you;
seek, ad ye shall find; knock, and it
shall be opened unto you.*

Mathew 7:7

Chapter 13 - Concept Visualization

Concept visualization is a helpful tool in building and understanding the why of a project before you address the what and how. It gives the client the opportunity to visualize the end-state and communicate the needs that must be met to ensure success.

The first time I saw concept visualization was back in about 1982.

I was working for Grumman Aerospace. I had a mentor who was a seasoned system engineer, he was talking to me about how he had mapped the vision for a command-and-control system and coupled that with the systems that were being built that were necessary to provide the inputs to the displays.

Those inputs could come from a collection of sensors, databases, and algorithms.

He said, let's go over to the archives and I will show you how we used to do it in the seventies.

We drove over to one of the plant buildings and went down into the basement.

He asked the archivist for a particular design package. The archivist came back to the role of brown butcher block paper. We rolled out the paper on a 30-foot drafting table.

The drawing had a series of displays that represented the information that the stakeholder needed to complete his or her job at any point of time.

Remember the discussion on how right-brain thinkers like to visualize? The team was able to show the user what the solution looked like before building the system.

Once Upon a Time, Visualization Was a Very Manual Process

He looked at me and said, this is how we used to do it.

The team sat down with the stakeholders who had to use those screens and visualize what it was going to be. Then they brought on the system views to figure out how in the hell they were going to feed that data in to give that view to that commander that was about to enter a battle or about to execute some other mission.

Asking WHY, asking stakeholders what they need, what benefits they expect, is not a new idea.

Some of this has been formalized and it's easier now than it was back in the seventies and eighties. As I said, this approach continues to be utilized today. Fortunately, today, we do not need to use a roll of butcher block paper, but it worked quite well in the past.

We need to know the "WHY". This is not a new idea. Understanding the Need to Know "Why" is Not New.

"He who has a "why" to live for can bear almost any how."

Friedrich Nietzsche (1888) "Twilight of the idols"

"Until thought is linked with purpose there is no intelligent accomplishment."

James Allen, circa 1903. "As a Man Thinketh"

"He [who] knows the "why" for his existence... will be able to bear almost any 'how'."

Viktor Frankl (1959) in *Man's Search for Meaning*

"Until we understand the 'why' of a requirement, we do not understand the requirement."

Jim Allen, circa 2018

I have been around a while, but Reverend James Allen from 1903 is a different individual and there is no relationship.

As you build the understanding of the "Why" of the project, you can move forward with the requirements that bridge the gap between the "why" and the ultimate solution.

Unfortunately, the pursuit of the ultimate solution often results in the ultimate personal struggle.

When I have something that causes me concern, I just dismiss everything connected with it from my mind and let my work absorb me, it's surprising how it clears up.

Henry Ward Beecher

Chapter 14 - The Ultimate Struggle

Intellectually, we may believe that we know that delivering value is the key to a successful program, but why is our role as a project manager such a struggle? The main reason is that our left brain is goal driven. Others measure us by hitting goals, not by delivering value. If you are like many, you strive to meet goals. That is how you get a bonus or a promotion.

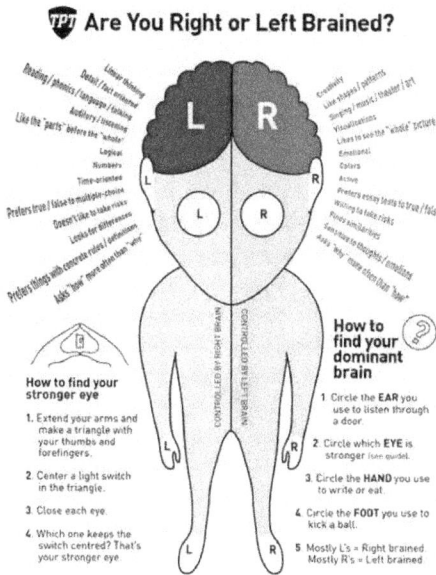

TPT Are You Right or Left Brained?

Fortunately, we each want to create value and be recognized for the value created. Our right brain dominates that part of our psyche. It is oriented toward being purpose driven. Do we need a craniology expert to guide us through life? Do left brain thinkers have a different skull shape?

Well, if you are like me, you are likely a left-brain thinker. I'm a Georgia Tech system engineer. I tried to build up the right side of my brain by getting a master's degree at the University of Virginia, but I still tilt to the left.

UVa tried to teach me some couth and made me start thinking with my right brain a little bit, but I found that you do need both sides of the brain.

Why do we need both sides?

Robert Shmerling, MD makes these observations in an interesting paper regarding the difference between right and left-brain thinkers.

Robert H. Shmerling, MD November 8, 2019

Right-brained thinkers are supposed to be intuitive and creative free thinkers. They are "qualitative," big-picture thinkers who experience the world in terms that are descriptive or subjective. While Left-brained people tend to be more quantitative and analytical. They pay attention to details and are ruled by logic.

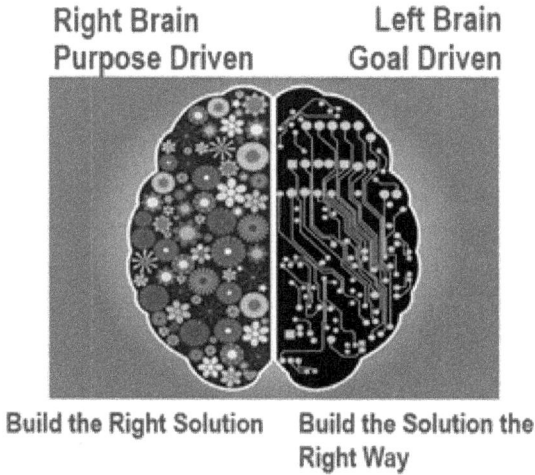

Right Brain
Purpose Driven

Left Brain
Goal Driven

Build the Right Solution

Build the Solution the
Right Way

I am more oriented toward the short-term schedule focused approach. I guess that you would say that I tend to be left brained. I have had to learn to compromise and use both sides.

Our left brain focuses on the near term. It is goal oriented and depends on short-term tactical thinking.

How does your left brain look at problems? How does it work?

What was I thinking in my left-brain?

- I wanted to be quantitative and internally focused.
- I have managed my time. I've allocated the time that was needed.
- I have a scarcity of time and ability to go through and do what I need to do.
- I must develop a quality presentation for this project, but I must be task oriented.
- I must collect data.

- I must go back and reread a lot of references that I've read in the past.
- I must pick excerpts out that show that this is not just Jim Allen's thinking.
- I'm paying my own bill for this and I'm not getting compensated.
- I was trying to minimize costs wherever possible.

Left-brain thinkers are very focused. Most left-brainers are very goal oriented. That is the way most left-brainers think and react.

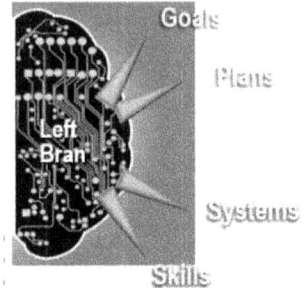

- Quantitative
- Internally Focused
- Position of Scarcity
- Task Oriented
- Revenue Focused

Hopefully we now understand that "Purpose" is very important to success, and we must learn to engage our right-brain. We need to look at the big picture. We must understand how our project will affect the ecosystem where it will reside.

We need to look beyond the initial requirements.

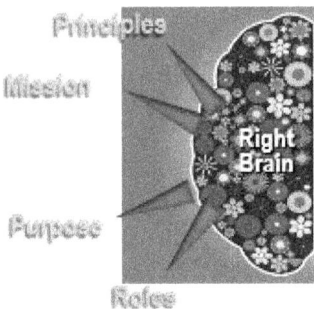

- Qualitative
- Externally Focused
- Position of Abundance
- Lead and Manage

- Creative
- Long Term Business Oriented

It is much easier to be purpose driven. Your right brain steps in and introduces other thoughts.

- I want to be qualitative in thinking about you and what I am saying to you?
- Is what I am saying important?
- I have a position to discuss, but I can go anywhere I want to go.
- I can talk about whatever I want to talk about, but I must make sure that I address your needs.
- I wish we were sitting around a table or in a conference room so that we could talk about this and share ideas.
- I would prefer to be more of a facilitator rather than a presenter.
- I must be somewhat creative.
- I must be long-term, business oriented.
- I must ask, "Is this going to help?"

Purpose is a better way of thinking in program management.

Be Purpose-Driven

"Purpose is not an initiative; it is a way of business. It must be core to the decisions, conversations, and behaviors across all levels to be

authentic and deliver the wealth of advantages it promises."

Jim Allen
Purpose@ResponseResource.com
www.ResponseResource.com

All growth depends upon activity. There is no development physically or intellectually without effort, and effort means work. Work is not a curse; it is the prerogative of intelligence, the only means to manhood and the measure of civilization.

Calvin Coolidge

Chapter 15 - Conclusion

As you progress in your project professional career, you must remember that the value that you provide your client defines your purpose. You must also remember that value can only be defined by the receiver, the client, not you.

Management may measure you on your ability to stay on budget and on schedule. That is fair. They are paying your salary. However, the client measures your success or failure using a different metric. The client measures you by the delivered value that you create. As a principled professional, fulfilling client needs and delivering value should be your paramount objectives.

Remember what PWC found, knowing that clients saw value in your work is the greatest motivator.

Creating value can be difficult. You may start by fulfilling a simple set of requirements but when the work is completed, the client says: "that is not what I needed." That outcome can only be overcome by understanding the "why" of the project from the stakeholder's perspective. Not from your own bias and usually not from the client's purchasing organization.

Establishing true needs requires collaboration;

collaboration requires trust. Building that trust requires that you establish the ability to bond with the client and see issues from the client's point of view. As I mentioned earlier, it is not enough to believe that you have the ability to be a surrogate client. You are the builder, not the user. Don't attempt to be the client.

You need to understand that hidden issues will not remain hidden. When the hidden issues are eventually exposed, clients are not likely to come forward and claim responsibility. You are the project professional. It is your responsibility to do what is right for the customer, not just to stay on budget and on schedule.

As Longfellow says in "The Builder",

> *Think not, because no man sees,*
> *Such things will remain unseen.*

Expose the underlying needs, the underlying requirements, the underlying client pain, and resolve each. That is your purpose, "the resolution of the underlying pain."

Please feel free to reach out to me. Get in touch if you have any specific problem or any other areas where you have an interest. I will be more than happy to work with you and share the lessons that I have learned over the past five decades.

We are considering offering this material as an

online course. If you are interested, leave a comment, at Purpose@ResponseResource.com and I will keep you informed.

Keep purpose ahead of goals and the goals will take care of themselves.

The secret to having a fulfilling job is to go home each day and realize that you provided value that was appreciated.

See everything from the other person's point of view.

To succeed, jump as quickly at opportunities as you do at conclusions.

Benjamin Franklin

References

Here are some of the references that I used in developing this book. You will find many more references that cover this subject.

Books:

Adams, Scott. *"Dilbert 2.0: 20 Years of Dilbert"* (2008)

Allen, David. *"Getting Things Done"* (2015)

Allen, James. *"As a Man Thinketh"* (1902)

Berne, Eric, M.D. *"Games People Play"* (1964)

Blanchard, Ken and Miller, Mark. *"The Secret"* (2001)

Brown, Rex V., Kahr, Andrew S. Peterson, Cameron.

"Decision Analysis: An Overview"

Carnegie, Dale. *"How to Win Friends and Influence People"* (1936) and *"Dale Carnegies Scrapbook"* (1959)

Cortucci, Robert. *"Competitive Intelligence: From Information to Action"* (2013)

DoD Deputy Chief Information Officer. *"DoD Architecture Framework Version 2.02"* (2010) https://dodcio.defense.gov/Library/DoD-Architecture-Framework/

Dweck, Carol S, Ph.D. *"Mindset the New Psychology of Success"* (2006)

Frankl, Victor and Winslade, William. *"Man's Search for Meaning"* (2006)

Gaffney, Steven. *"Honesty Works!"* (2006)

Grusin, Jay, PhD with Lindo, Steve. *"Intelligent Analysis"* (2021)

Harris, Thomas A, M.D. *"I'm Ok You're Ok"* (1969)

Kendrick, Tom. *"How to Manage Complex Programs"*, AMACOM, (2016)

Kerth, Norman. *"Project Retrospectives"* (2001)

Kolenda, Nick, Methods of Persuasion: How to Use Psychology to Influence Human Behavior" (2013)

Krupp, Steven and Schoemaker, Paul. *"Winning the*

Long Game" (2014)

Nietzche, Fredrich. *"Twilight of the Idols"* (1889)

Olsen, Dan. *"The Lean Project Playbook"* (2015)

Project Management Institute. *"A Guide to the Project Management Body of Knowledge"* (PMBOK Guide – 2000 Edition)

"A Guide to the Project Management Body of Knowledge" (PMBOK Guide – Fifth Edition)

"A Guide to the Project Management Body of Knowledge" (PMBOK Guide – Sixth Edition)

"Agile Practice Guide" (2017)

" Business Analysis for Practitioners" (2015)

Reiman, Joey. *"The Story of Purpose"* (2012)

Roden, Tom and Williams, Ben. *"Fifty Quick Ideas to Improve Retrospectives"* (2015)

Scheessele, William. *"Winning Conversations"* (2010)

Sinek, Simon; Mead, David; Docker, Peter. *"Find Your Why: A practical Guide for Discovering Purpose for You and Your Team"* (2017)

Steiner, Claude. *"Scripts People Live"* (1974)

Walker, Gordon. *"Modern Competitive Strategy"* (2007)

Articles:

Hosselney, Majeed. *"The Hard and Soft Components of Purpose-Driven Project Leadership"* (2021) https://www.forbes.com/sites/forbesbusinessdevelopmentcouncil/2021/01/12/the-hard-and-soft-components-of-purpose-driven-project-leadership/?sh=136396ec55a8

Kruger, James. "Mastering the Art of Deal-Making: A Paradigm Shift to Client-Centr4icity (2023)

https://www.linkedin.com/pulse/mastering-art-deal-making-paradigm-shift-james-Kruger/

Microsoft CoPilot AI. *"Purpose-driven project management"* (2024)

Search on Edge

O'Donovan, Caroline. *"Changes at Amazon-owned health services cause alarm among patients and employees"* (Washington Post March 1, 2024)

Nieto, Antonio, *"Does Your Project Have a Purpose?"* (2021) https://hbr.org/2021/10/does-your-project-have-a-purpose

Project Management.com. *"The Project Manifesto"* (2024)

https://www.projectmanagement.com/articles/937468/the-project-manifesto--12-guiding-principles-of-the-project-economy

Perrine, Lisa Ed. D. *"Purpose-Driven Projects: Start with Why"* (2019)

    https://www.avixa.org/pro-av-trends/articles/purpose-driven-projects-start-with-the-why

Proffitt, Vince. "The Power of Putting Purpose Before Profit" (2023) https://www.inc.com/inc-masters/the-power-of-putting-purpose-before-profit.html

Ryan, Mike. *"The Role of Stakeholders in Requirements Elicitation"* (2014) https://incose.onlinelibrary.wiley.com/doi/10.1002/j.2334-5837.2014.tb03131.x

    "On the Format of Project Purpose Statement" (2011) https://www.researchgate.net/publication/270219635_On_the_Format_of_a_Project_Purpose_Statement

Shannon Schuyler, PWC. *"Putting Purpose to Work"* https://www.pwc.com/us/en/purpose-workplace-study.html

.

For every ailment under the sun,
There is a remedy, or there is none;
If there be one, try to find it;
If there be none, never mind it.

Chinese Proverb

Glossary of Terms

ARDAK Corporation – Competitive pricing source

BOE - Basis of Estimate

BOM - Bill of Material

BRM - Benefits Realization Management

BU – Business Unit

C3 – Triple constraints: schedule, scope, and budget

CPARS – Contractor Performance Assessment Reporting System

CPI – Cost Performance Index

CV - Capability View

DoD - Defense Department

DoDAF - DoD Architectural Framework

FOIA - Freedom of Information Act

MAU - Multi-Attribute Utility Analysis

PC – Pain of Change

PM – Project Manager

PMBOK – Project Management Body of Knowledge

PMI – Project Management Institute

PMO – Project Management Office

PMP – Project Management Plan

PRR - Proposal Readiness Review

PTI – Pain They are In

PVM – Project Value Management

PWC – Price Waterhouse Cooper

RACI - Responsible, Accountable, Consulted, and Informed

RFP – Request for Proposal

ROI – Return on Investment

SME - Subject Matter Expert

SOW – Statement of Work

SV - System View

SWOT - Strengths, Weaknesses, Opportunities, and Threats

TRR - Technical Readiness Review

VPC – Value/Price Comparison

WIIFM – What's in it for me?

WOTS – Weaknesses, Opportunities, Threats, and Strengths

Appendix A – Sample Evaluation Factors

(Extracted from Contract - N0017324RBZ01)

As you do your self-scoring, try to use the clients' factors for evaluation and their criteria for scoring each of those factors. The following is a good example of Section M - Evaluation Factors for Award extracted from an actual competition. Remember, look at offers from the client's perspective, not yours.

If you do not have section M, do your research, and find one that the client previously published. Many clients tend to reuse.

M-1 BASIS OF AWARD

The Government intends to make an award to that responsible vendor whose proposal is determined to be the best overall value to the Government in accordance with the established criteria and rating methodology set forth in the RFP.

The evaluation will be based on a complete assessment of each Offerors technical proposal, past performance information, and cost proposal. Within the best value continuum, the Government will employ a Price/Technical Tradeoff analysis of both price and non-price factors (FAR 15.101-1) in evaluating each proposal submitted. Trade-off considerations may result in the determination that it is in the best interest of the Government to award someone other than the lowest priced Offeror or other than the highest technically rated Offeror. A best value analysis will not be performed or developed for any Offeror whose proposal is found to be technically unacceptable or unsatisfactory in any other factor or subfactor.

All proposals shall be subject to evaluation by a team of Government personnel.

Proposals will be assessed on how well each Offerors proposal meets the solicitation requirements and the risks associated with the Offerors approach. Determining how well the Offerors proposal meets the solicitation requirements will be accomplished in two steps. First, a determination will be made if the Offerors proposal meets the solicitation requirements. Next, the discriminators will be identified for the proposals reflecting the unique strengths, weaknesses, significant weaknesses, and deficiencies of each offer. In addition, the Government will examine the benefit or risk to contract performance of each discriminator and assess its relative value to the Government. To make a sound selection decision, the Government needs to understand the ways in which a given proposal is considered technically strong, as well as the ways in which it is weak or deficient.

Therefore, the Government will evaluate the identified strengths, weaknesses, and deficiencies (in terms of the evaluation criteria) to facilitate the process of determining which proposal presents the best overall value to the Government.

The Offerors will receive one overall rating value for the non-price proposal evaluation factors. Price, while being an important factor, is not in and of itself the determining factor in the selection of the successful Offeror for award of the contract contemplated by this solicitation. The Contracting Officer may reasonably determine that the superior personnel and staffing ability, technical understanding, and capability, and/or management capability merits a higher price, and therefore represents the best value to the Government. The Contracting Officer, using sound business judgment, will base the selection decision on the integrated assessment of the Offeror(s)' non-price factors and price factor measured against the evaluation criteria listed under section M-2 below.

Award will only be made to an Offeror that has no organizational conflict of interest as defined in FAR 9.5 or that the Government determines has provided a satisfactory mitigation plan. Offerors are advised that technical proposals may be evaluated without consideration of any proposed Subcontractor which is deemed to have an organizational conflict of interest and for which an unsatisfactory mitigation plan has been proposed. Failure by an Offeror that has identified a potential OCI or to submit an OCI mitigation plan with its proposal shall no longer being considered for award.

M-2 EVALUATION FACTORS FOR AWARD

The Government will evaluate each Offerors proposal in accordance with the factors contained in Section L and listed below to determine the best value proposal. The evaluation factors represent key areas of importance to be considered in the source selection decision. The factors and associated elements have been chosen to support meaningful discrimination between and among competing proposals. As demonstrated in each proposal, a prospective Offeror shall be evaluated in terms of its ability to meet or exceed the task requirements stated in the IDIQ SOW.

Each proposal shall be evaluated in accordance with the factors listed in the table below.

EVALUATION FACTORS
F actor 1 - Technical Understanding, Capability, and Approach
Factor 2 - Management Capability
Factor 3 - Staffing and Personnel Qualifications
Factor 4 - Past Performance

Factor 5 - Cost/Price

Order of Importance:

Factor 1 - Technical Understanding, Capability, and Approach is more important than Factor 2 - Management Capability, and Factor 3 - Personnel Qualifications. Factor 4 - Past Performance is more important than Factor 5 - Cost/Price.

Factor 2 - Management Capability, and Factor 3 - Personnel Qualifications are equally important.

Factor 1 - Technical Understanding, Capability, and Approach, Factor 2 - Management Capability, and Factor 3 - Personnel Qualifications are each more important than Factor 4 - Past Performance, and Factor 5 - Cost/Price.

All evaluation factors other than cost or price, when combined, are significantly more important than Factor 5, Cost/Price. However, as non-price factors become closer in perceived value, price considerations will become more important. Trade-off considerations may result in the determination that it is in the best interest of the Government to award to someone other than the lowest priced Offeror or other than the highest technically rated Offeror.

To be considered for award, a rating of no less than "Acceptable" must be achieved for Factors 1 - 3. An

"Unacceptable" evaluation rating for any factor will result in an "Unacceptable" Factor and will render the Offeror ineligible for award.

M-3 EVALUATION OF FACTORS

FACTOR 1 - TECHNICAL UNDERSTANDING, CAPABILITY, AND APPROACH

The Government will evaluate degree to which the Offerors technical proposal clearly demonstrates the capability, knowledge, and approach, for both Prime and Subcontractors in performing all aspects of the IDIQ SOW. The degree to which the Offeror demonstrates an overall understanding of the scope of work and provides an approach to performing the tasks described in the RFP with particular emphasis on projects with scientific, engineering, and technical tasks similar in size, scope, and complexity to those required in the IDIQ SoW. The documentation should be sufficient to support both the prime and subcontractors' breadth and depth of experience as it relates to the IDIQ SoW and should clearly demonstrate:

(1) The relationship between the company's experience and the tasks required under the IDIQ SoW
(2) Prior or current programs in the task areas and
(3) Key/critical aspects and challenges associated the task areas with a strategy to address/mitigate.
(4) Technical understanding of all task areas.
(5) Feasibility of approach

FACTOR 2 - MANAGEMENT CAPABILITY

The proposal will be evaluated on the Offerors demonstrated management capability and success in managing projects of similar complexity and duration as that set forth in the IDIQ SOW. The Government will evaluate the extent to which the Offerors proposed approach for managing the contract effort, and its organizational structure, internal management, and communications processes/tools, will enable successful performance of the IDIQ SoW requirements.

The Government will evaluate whether the Offeror presents a management plan that provides an integrated team with a coordinated approach to work performance, demonstrates a clear understanding of solicitation requirements, assures quality long term support, and demonstrates the Offerors management ability and success in managing projects of similar size, scope, and complexity as that set forth in the IDIQ SOW.

FACTOR 3 - STAFFING AND PERSONNEL QUALIFICATIONS

Staffing and Personnel Qualifications will be evaluated on the Offeror's ability to provide personnel in accordance with the Personnel Qualifications document to carry out the Task Order

SOW. For non-key personnel, the Offeror will be evaluated on its approach and ability to provide or to recruit, retain and train adequate staff with the requisite skill sets and security clearances to meet the requirements of the Task Order SOW by the start of order period of performance.

If proposed, the Government will evaluate the subcontracting team breakout including the teaming partner's name; CAGE Code and associated active SAM registration, if applicable; major/minor subcontractor distinction; subcontract type and any additional risk this may bring to the Government; and the % of Total Labor Hours Performed by the subcontractor in correlation to the overall total hours proposed by the Offeror.

If major cost reimbursement subcontractors do not comply with providing cost data at the same level of detail as the Prime, this may be viewed as indicative of the Prime contractor's inability to manage Subcontractor performance and may impact the Staffing and Personnel Qualifications evaluation result.

The proposal will be evaluated on the availability of proposed key personnel to support the effort, and the extent to which the proposed key personnel satisfy the minimum required qualifications for key personnel for the first task order only. The evaluation will consider any qualifications that exceed the minimum qualifications (e.g., greater education, additional years of experience, relevant experiences, certifications, etc....) and whether those qualifications will benefit the Government.

The Government will evaluate the degree to which required resumes demonstrate the Offeror's knowledge and ability to successfully meet requirements of the Task Order SOW and related competencies, the relevant experience the proposed personnel have in performing each of the Task Order SOW areas, the level of the personnel's relevant education and training, the personnel's security clearances, and the overall quality of key personnel proposed (resumes).

The Government will evaluate the Offeror's demonstrated ability to provide specific personnel to this effort. Failure to submit resumes, Subcontract agreements, or LOI(s) for key personnel, as required, will result in an unacceptable rating.

Key personnel, as identified in this solicitation, shall currently be employed by the Offeror or documentation included showing their immediate availability. A statement of commitment by the Offeror that specific personnel will be committed to the effort is essential, and the amount of effort each will be performing against the Task Order. Failure to submit the appropriate documentation showing the availability of the proposed key personnel may result in an unacceptable rating, rendering the proposal not awardable.

FACTOR 4 - PAST PERFORMANCE

Past performance is a measure of the degree to which the Offeror satisfied its customers in previous relevant contracts and complied with Federal, State, and local laws and regulations. Each Offerors (and major subcontractor, team partner, etc., if applicable) past performance will be reviewed to determine recency, relevancy, and the quality of performance on the past contracts. A past performance confidence assessment rating will be assigned based on the Offerors overall record of recency, relevancy, and quality of performance.

The Government will evaluate submitted CPARS reports, Past Performance Questionnaires and Previous Contract Effort Narratives, and may contact the Offerors customers to ask whether or not they believe: (1) that the Offeror is capable, efficient and effective; (2) that the Offerors performance conformed to the terms and conditions of its contract; (3) that the Offeror was reasonable and cooperative during performance; (4) that the Offeror was committed to customer satisfaction; and (5) if given a chance would they select the same or a different Contractor.

The Government may consider past performance information obtained from sources other than those identified by the Offeror to evaluate an Offerors or subcontractors past performance, including Federal, State, and local Government agencies, Better Business Bureaus, published media and electronic databases. The lack of recent and relevant past performance information will result in the assignment of a neutral rating (i.e. neither favorable nor unfavorable) for this factor. The Government reserves the right to limit or expand the number of references it decides to contact and to contact other references than those provided by the Offeror or subcontractors.

FACTOR 5 - COST/PRICE

Cost/Price evaluation will be based on an analysis of the completeness of the cost and pricing data, realism, and reasonableness. Cost Completeness means the proposed costs are adequately identified, estimated, and supported.

Cost Realism (FAR 2.101) means the proposed costs are (1) realistic for the work to be performed, (2) reflect a clear understanding of the requirements, and (3) are consistent with the various elements of the Offerors technical proposal. Cost reasonableness (FAR 31.201) is defined as reasonable if the cost does not exceed the amount incurred by a prudent person in the conduct of a competitive business. The overall evaluated price, inclusive of base and all options, shall be evaluated to ensure that it is fair and reasonable.

If proposed, the Government will evaluate the Prime's subcontracting teaming arrangement and each major subcontractor's sanitized cost proposal for fair and reasonableness. In accordance with FAR 15.404-3(b), the Offeror shall conduct appropriate price analysis to establish the reasonableness of each proposed subcontractor's price. The Government will evaluate how the Prime determined price reasonableness for each subcontractor proposed.

Appendix B – CEP Course Summary

The following summary schedule is for my course on client engagement that was previously approved for seven PMI continuing education units. The course is not currently offered but much of the material can be found in this book.

Contact *Purpose@ResponseResource.com* if you would like to discuss additional training opportunities.

Recommended one-day class schedule and alignment with the PMBOK 5ᵗʰ Edition and PMI Talent Triangle

Module	ACTIVITY	Duration	PMBOK Reference 5th Ed.
	Class starts promptly at 8:00		
	INTRODUCTIONS	10	
1	SHAPING AND TRACKING RELATIONSHIPS AND OPPORTUNITIES	20	4.4.2, 10.1.2, 13.1.2, X3.1, X3.2

2	INTRODUCTION TO THE HUMINT CEP	5	10.1.2
3	HOMEWORK BEFORE THE CALL	75	10.1.2, 13.2.2, 13.4.1.4, X3.7
4	PREPARING FOR THE CALL	45	10.2.2, 13.3.2
	Break	15	
5	PURPOSE AND GOAL	40	10.3.2, 13.3.2
	NEGOTIATING RULES, RIGHTS & RESPONSIBILITIES	15	10.3.2, 13.3.2, X3.9
	EXECUTING THE CALL	15	10.3.2, 13.3.2, X3.4
	BONDING AND POSITIONING	45	10.3.2, 13.3.2, X3.9
	Lunch	60	
6	THE ART OF THE INTERVIEW	15	10.3.2, X3.9
	INTRO TO INTEL GATHERING SKILLS	10	10.3.2, X3.4
	INTERVIEWING AND QUALIFYING SKILLS	15	10.3.2, X3.5
	NEWTONS LAWS & NEGATIVE REVERSING	10	10.3.2. X3.4

	THE DUMMY PROCESS & SKILLS	15	10.3.2, X3.4
7	ASKING PERMISSION TO INTERVIEW	15	10.3.2, x3.8
	Break	15	
	THE PAIN INTERVIEW	20	10.3.2, X3.8
	CLOSING THE CALL	10	10.3.2
8	QUALIFYING FOR FINANCIAL ABILITY	15	13.3.2, X3.8
	QUALIFYING FOR DECISION MAKING PROCESS	20	13.3.2, X3.6
	WHEN AND HOW TO GIVE A PRESENTATION	10	13.3.2
	LAUNCHING A PRESENTATION	10	13.3.2
9	POST CALL PHASE	15	4.4.2, 13.3.3.5, 13.4.2.1
	Total Time	540	
	Intro., Break and Lunch Time	100	
	Class Time	440	
	Suggested PDUs	7	

*Life does not consist mainly – or
even largely – of facts and
happenings. It consists mainly of
the storm of thoughts that is forever
blowing through one's head.*

Mark Twain

Appendix C – Value Proposition

My website provides a reusable deck that I find useful in team exercises to build valid Value Propositions. Please feel free to download, modify, and use as you deem appropriate.

ResponseResource.com (select the tools option)

If you follow the template of the deck, you could effectively spend several hours building your value proposition. List of the slides in the deck:

Value Questions:

- How do we determine value?
- What are the value drivers?
- What forces effect value?
- What does the customer value?
- What does the customer know about us?
- What does the customer state are his needs?
- What does the customer infer that he wants?
- How can we prove our capability?
- How will the customer benefit (WIIFM)
- A Value Proposition Template
- Strategy, What and Why
- What kinds of questions need to be answered?

Management Decision Questions

- How can we win without a strategy?
- How do we develop a successful strategy?

- How do we achieve a profitable market position?
- How do we defend our position?
- How do we build a competitive advantage?
- Why should we pursue uniqueness?
- Why should activities align?
- Why should we choose what not to do?
- Where does Operational Innovation fit in?
- Is price competition inevitable?

Strategic Analysis

- Lumpy and Risky Considerations
- Strategy is a Process

Forces that effect Value, Price, and Cost

- Competition
- Barriers to entry
- Buyer Power
- Supplier Power
- Substitutes

Appendix D – Selected DoDAF Views

DoDAF (DoD Architectural Framework) provide a useful approach to defining the capabilities of a system and the service that comprise that system.

There are multiple views that you might find helpful as project managers. I include the operational, services and capabilities model descriptions.

Operational Model Descriptions

OV-1: High-Level Operational Concept Graphic	The high-level graphical/textual description of the operational concept.
OV-2: Operational Resource Flow Description	A description of the Resource Flows exchanged between operational activities.
OV-3: Operational Resource Flow Matrix	A description of the resources exchanged and the relevant attributes of the exchanges.
OV-4: Organizational Relationships Chart	The organizational context, role, or other relationships among organizations.
OV-5a: Operational Activity Decomposition Tree	The capabilities and activities (operational activities) organized in a

	hierarchal structure.
OV-5b: Operational Activity Model	The context of capabilities and activities (operational activities) and their relationships among activities, inputs, and outputs; Additional data can show cost, performers, or other pertinent information.
OV-6a: Operational Rules Model	One of three models is used to describe activity (operational activity). It identifies business rules that constrain operations.
OV-6b: State Transition Description	One of three models is used to describe operational activity (activity). It identifies business process (activity) responses to events (usually, very short activities).
OV-6c: Event-Trace Description	One of three models is used to describe activity (operational activity). It traces actions in a scenario or sequence of events.

Service Model Descriptions

Model	Description
SvcV-1 Services Context Description	The identification of services, service items, and their interconnections.
SvcV-2 Services Resource Flow Description	A description of Resource Flows exchanged between services.
SvcV-3a Systems-Services Matrix	The relationships among or between systems and services in a given Architectural Description.
SvcV-3b Services-Services Matrix	The relationships among services in a given Architectural Description. It can be designed to show relationships of interest, (e.g., service-type interfaces, planned vs. existing interfaces).
SvcV-4 Services Functionality Description	The functions performed by services and the service data flows among service functions (activities).
SvcV-5 Operational Activity to Services Traceability Matrix	A mapping of services (activities) back to operational activities (activities).
SvcV-6 Services Resource Flow Matrix	It provides details of service Resource Flow elements being exchanged between services and the attributes of that exchange.
SvcV-7 Services Measures Matrix	The measures (metrics) of Services Model elements for the appropriate timeframe(s).

SvcV-8 Services Evolution Description	The planned incremental steps toward migrating a suite of services to a more efficient suite or toward evolving current services to a future implementation.
SvcV-9 Services Technology & Skills Forecast	The emerging technologies, software/hardware products, and skills that are expected to be available in a given set of time frames and that will affect future service development.
SvcV-10a Services Rules Model	One of three models is used to describe service functionality. It identifies constraints that are imposed on systems functionality due to some aspect of system design or implementation.
SvcV-10b Services State Transition Description	One of three models is used to describe service functionality. It identifies responses of services to events.
SvcV-10c Services Event-Trace Description	One of three models is used to describe service functionality. It identifies service-specific refinements of critical sequences of events described in the Operational Viewpoint.

Capability Model Descriptions

Model	Description
CV-1: Vision	Addresses the enterprise concerns associated with the overall vision for transformational endeavors and thus defines the strategic context for a group of capabilities.
CV-2: Capability Taxonomy	Captures capability taxonomies. The model presents a hierarchy of capabilities. These capabilities may be presented in context of a timeline - i.e., it can show the required capabilities for current and future capabilities.
CV-3: Capability Phasing	The planned achievement of capability at different points in time or during specific periods of time. The CV-3 shows the capability phasing in terms of the activities, conditions, desired effects, rules complied with, resource consumption and production, and measures, without regard to the performer and location solutions
CV-4: Capability Dependencies	The dependencies between planned capabilities and the definition of logical groupings of capabilities.
CV-5: Capability to Organizational Development Mapping	The fulfillment of capability requirements shows the planned capability deployment and interconnection for a particular Capability Phase. The CV-5 shows the

	planned solution for the phase in terms of performers and locations and their associated concepts.
CV-6: Capability to Operational Activities Mapping	A mapping between the capabilities required and the operational activities that those capabilities support.
CV-7: Capability to Services Mapping	A mapping between the capabilities and the services that these capabilities enable.

Appendix E – Builders – aka Project Managers

Project managers are builders of solutions. "Our to-days and yesterdays are blocks with which we build." Take pride in your work and as Longfellow says, "Leave no yawning gaps between."

"The Builders"
By Henry Wadsworth Longfellow

All are architects of Fate,
Working in these walls of Time;
Some with massive deeds and great,
Some with ornaments of rhyme.

Nothing useless is, or low;
Each thing in its place is best;
And what seems but idle show
Strengthens and supports the rest.

For the structure that we raise,
Time is with materials filled;
Our to-days and yesterdays
Are the blocks with which we build.

Truly shape and fashion these;
Leave no yawning gaps between;

Think not, because no man sees,
Such things will remain unseen.

In the elder days of Art,
Builders wrought with greatest care
Each minute and unseen part;
For the Gods see everywhere.

Let us do our work as well,
Both the unseen and the seen;
Make the house, where Gods may dwell,
Beautiful, entire, and clean.

Else our lives are incomplete,
Standing in these walls of Time,
Broken stairways, where the feet
Stumble as they seek to climb.

Build to-day, then, strong, and sure,
With a firm and ample base;
And ascending and secure
Shall tomorrow find its place.

Thus alone can we attain
To those turrets, where the eye
Sees the world as one vast plain,
And one boundless reach of sky.

Face your deficiencies and acknowledge them; but do not let them master you. Let them teach you patience, sweetness, and insight. True education combines intellect, beauty, goodness, and the greatest of these is goodness. When we do the best that we can, we never know what miracle is wrought in our life, or in the life of another.

Helen Keller

Response Resource Publication

www.ingramcontent.com/pod-product-compliance
Lightning Source LLC
Chambersburg PA
CBHW071648200326
41519CB00012BA/2441